TURBOSPACE

JASON O'NEIL

authorHOUSE®

AuthorHouse™
1663 Liberty Drive
Bloomington, IN 47403
www.authorhouse.com
Phone: 1 (800) 839-8640

This is a novel about space and a potential mission using an anti-gravity device yet to exist.

The names of corporate and government entities, products, and/or persons are not an endorsement by the author in any manner and should be considered coincidental.

Published by AuthorHouse 05/16/2016

ISBN: 978-1-5246-0882-8 (sc)
ISBN: 978-1-5246-0880-4 (hc)
ISBN: 978-1-5246-0881-1 (e)

Library of Congress Control Number: 2016907757

Print information available on the last page.

This book is printed on acid-free paper.

CONTENTS

PREFACE

In the mid-1600s it took Isaac Newton decades to comprehend universal gravitation. To accomplish this Newton made many connections to other realms of thought. He used a wide variety of sources to determine the speed of the Earth. He struggled to determine the rate of fall due to gravity using calculations made by Galilei Galileo. In the end he concluded that a body on the Earth's surface is drawn downward by gravity 350 times stronger than the Earth's inclination to fling it outward into space.

Newton estimated that Earth attracted an apple 4,000 times more powerfully than it attracted the moon using an inverse-square law applied to observations by Johannes Kepler. He is credited with developing the principles of motion and force, including "A body once moved will always keep the same celerity, quantity and determination of its motion." (Opticks, Page 404)

Early in the 21st century, with technology evolving approximately every 18 months, a novel Red Box is invented which neutralizes the downward force of gravity. It is applied in a novel, personal, all-electric ground and air vehicle called a Turbopod, or "T-Pod. Because of the Red Box, no rocket is required to achieve escape velocity for a space capsule to orbit the Earth, and with only a little thrust it can be sent to the International Space Station, the moon, and into deep space should mankind decide it is worth the effort.

CAST OF CHARACTERS

MATT FLYNN
Principal, entrepreneur, always wanted to fly like a bird
Age: 58
Married to Heather
Good salesman, presenter
Medium build, athletic
Similar to: Burt Lancaster, actor

MURRAY FLYNN
Principal, physicist, inventor of the Red Box
Age: 55
Younger brother of Matt; married to Maggie
Astronomical math IQ; considered a genius
Needs Maggie's common sense
Similar to: Robert Oppenheimer; nuclear scientist

HEATHER FLYNN
Aeronautical Engineer
Age: 56; looks 46
Married to Matt
Very inquisitive; always asks "Why?"
Tall, slender, red hair and freckles
Helped design the Red Box
Similar to: Cintia Dicker, Brazilian model

MAGGIE FLYNN
Pilot, Skilled in martial arts
Age: 51
Married to Murray
Slender, athletic body, dark hair and features
Murray's alter-ego
Works well with Captain Russell
Similar to: Maggie Quinlan (Maggie Q), actress

JIM RUSSELL
Chief Pilot; Flew 747s for 18 years
Age: 58
Detailed-oriented; patient; good teacher
Trains Turbopod pilots
Similar to: Harrison Ford, actor

CHIN CHIN PO
Chinese detective based in Hong Kong
Age: 46
Tall, delicate features, narrow face
International reputation for solving crimes
Astronaut
Similar to: Li Bing Bing, Chinese actress and model

WERNER VON BOLTZ
Chairman, Mercedes-Benz; rose through the ranks
Age: 62
Attended Heidelberg University's Max Planck Institute for Geophysics
Private pilot
Similar to: Wernher von Braun, rocket engineer, designer

REINER STRASSBURG
Chief Engineer, Mercedes-Benz, Stuttgart, Germany
Age: 64
Considered a styling genius in the automotive industry
Similar to: Konrad Adenauer, former Chancellor of Germany

REINHOLD TIMM
Factory Manager, Mercedes-Benz, Vance, Alabama
Age: 56
Knows everything about vehicle manufacturing
Stern but fair; respected by employees and management
Avid bass fisherman
Similar to: John Glenn, astronaut

JENNINGS WERNER
First Independent party U.S. President
Age: 52
New Englander
Tall, imposing figure with large "political" hands
Successful lawyer turned legislator; master of backroom deals
Similar to: large John F. Kennedy, U.S. President

JEFFERSON PALMER
U.S Secretary of Defense
Age: 54
University of Virginia graduate
Retired U.S. Air Force Lieutenant General
Tall black man; former Thunderbird pilot
Similar to: Colin Powell, U.S. Secretary of State

DON GOLDEN
NASA Administrator
Age: 58
35-year career as aerospace engineer and corporate leader
Expert in American classified spacecraft
Similar to: Dan Golden, retired NASA Administrator

DR. CLIFF ARNOLD
NASA Chief Scientist
Age: 62
Caltech and MIT graduate
Worldwide reputation as propulsion engineer; rock scientist
Similar to: Dr. Gerard O'Neill, physics professor

MERRITT PRESTON
Space Company Launch Director, Boca Chica, Texas
Age: 56
Retired from NASA Kennedy Space Center
Designer of spacecraft Integration & Test facilities
Similar to: James Stewart, actor

ALTON TAYLOR
Space Express Chairman
Age: 58
Former U.S. Air Force test pilot and astronaut
Flight engineering; graduate of Embry-Riddle
Similar to: Elon Musk, automotive and space entrepreneur

KATIA RUSKOV
Astronaut, engineer, payload specialist
Age: 36
Karate black belt
Grew up near Russian Launch Complex
Similar to: Sally Ride, astronaut

HANS ULRICH
Astronaut, 3-D manufacturing expert, payload specialist
Age: 39
Mercedes-Benz engineer on leave to astronaut corps
Similar to: Bernhard Langer, professional golfer

LENNON BLAIR
Astronaut, propulsion expert, energy management engineer
Age: 42
Royal Air Force jet pilot
Similar to: Prince Harry of the United Kingdom

MIKO MIYAKI
Astronaut, robotics expert, payload specialist
Age: 31
Graduated first in her class in college in Japan
Author of mathematics books
Similar to: Nana Fujimoto, female hockey player

TOM "CATFISH" CROWLEY
Bald Eagle spacecraft commander
Age: 50
Senior member of the astronaut corps
Master of astronomy and star formations
Similar to: Ernest Hemingway

HERMAN "FISH" SALMON
U.S. Air Force test pilot who broke many speed records
Age: 51
Bald Eagle spacecraft pilot
Similar to: Richard Gere, actor

TRYG AGER
Nautical construction engineer
Age: 50
Teacher in Bergen, Norway; part-time consultant to marine industry
Similar to: Liam Neeson, actor

NILS NORGARD
Marine engineer; mechanical engineer
Age: 52
Former ocean liner designer
Expert in tidal and wave dynamics
Similar to: Eric Sevaride, journalist

DR. SUN WEI
Director, People's Aerospace Department, Beijing
Age: 60
Former space launch director
Leader, military spacecraft; member, China's moon mission project
Similar to: Chou En Lei, Chinese Diplomat

CLAUDIA SANTINI
Astronaut in Brazil; military aircraft pilot
Age: 39
Bald Eagle spacecraft pilot
Similar to: Jeísa Chiminazzo, fashion model

CHAPTER 1

COST OF SPACE

"Murray, now that you and the engineers at Mercedes-Benz have figured out a way to get to Mars and return safely, we owe NASA the proposal we promised," Matt Flynn told his brother. "The Red Box and Turbopod play key roles in the mission. Let's package your concept and all the calculations into a briefing for presentation at the end of the month. NASA has asked us to provide a bid for our consulting services, and we don't want to disappoint them, much less President Werner."

"I know, Matt, but …"

"But what?"

"Well the more I think about it, the more I realize that going to Mars is the wrong thing to do," Murray said.

"What do you mean?" Matt asked with a real concern in his voice.

"Well, brother, let's review the situation," Murray replied. "First, why are we going there? The answer can't be that it's in the human nature to explore. That was fine for Columbus. But his goal was to find riches for Spain. We can't say that about Mars. After 15 probes of one kind or another, we're not even sure there's water there, much less riches of some sort to benefit mankind. In my opinion, the return on the investment simply isn't there."

"Murray, please explain your logic to your dense brother," Matt said.

"OK. I'll try," Murray said. "First, let's look at the cost. Industry calculates that to put a kilogram — that's 2.2 pounds — in orbit costs $10,000. Each Apollo landing on the moon cost $18 billion in 1960s dollars. Given the number of necessary launches and infrastructure to support a Mars mission, the cost could be $500 billion dollars. Even a consortium of nations doesn't have that kind of money. And there are more pressing social needs today. But the real question — the show-stopper, really — is what do we get for the money?"

Matt looked intently at his brother, then took a deep breath and said, "I'm listening. What are your concerns?"

"Well let me list a few," Murray replied. "First, Mars' atmosphere is 1 percent of that of the Earth. We would have to bring our own oxygen, and to make enough oxygen to support a base would take many missions. And while Mars' daytime temperature is about 60 degrees, at night it's minus 100 degrees."
"That's pretty cold," Matt said.

"And that's not all," Murray continued. "You can't take enough water. One scientific estimate is 127 tons of water would be required to support a small team of astronauts during the 1,100-day mission," Murray said. "And they have to eat, too. Based upon our experience at the International Space Station, each astronaut would need 2 tons of freeze-dried food. Of course, they won't eat much if the sun's radiation kills them first. And the swelling of the optic nerve could lead to blindness."

"You're a real ray of sunshine, you know that?" Matt joked.

"I'm serious, Matt. The Martian soil may be toxic with its chlorinated salts would be bad for the thyroid gland. And dust would cover everything," Murray said. "Which is to say nothing of the psychological effects of such a mission. The recently completed Mars-500 project, where a small crew remained isolated in a small enclosure, highlighted the potential psychological problems that astronauts could face."

"What about mining Mars?" Matt asked.

"Mining would be a waste of time and resources," Murray said. "It wouldn't be cost-effective to bring it back. Don't you see, Matt? All mission planning to date is single-string. There is no redundancy to reduce risk. Even our own plan is single-string!"

"OK, I get the picture, Murray," Matt said. "If not Mars, what about Venus?"

"It's the same situation but even worse — it's 750 degrees there!" Murray exclaimed.

"Wow. I had no idea," Matt said, shaking his head from side to side. "So what's the solution?"

Murray was quick to reply. "In our briefing to NASA, we provide two missions. The first is the Mars mission we've developed, complete with its quarter-trillion-dollar price tag. The second mission is to the moon at a fraction of the cost and with the potential for a real payback. The plan is to develop technologies on the moon that can be directly applied to counter Earth's global warming."

"Like what?" Matt asked.

"Well," Murray began, "there are a number of technologies that can be developed and tested on the moon which can make a difference — an indoor greenhouse with hydroponics on a large-scale such as an O'Neill cylinder, economic desalination to split water to get oxygen and longer-life batteries that are truly disposable."

"Impressive," Matt said.

"And that's just the tip of the iceberg. There could be inflatable structures with hardened ceramic coatings, wearable sensor systems on clothing, advanced retinal communications, anti-muscle- and bone-loss processes, even lower cost solar power from 3-D printed panels. And there are dozens more we can discuss with the NASA engineers and scientists."

"I see what you mean, Murray," Matt said. "Saving our planet is a top priority, and the moon base can be instrumental in this effort. But what if NASA insists on going to Mars? Can we play an active role? Should we?"

Murray thought for a moment before replying. "Matt, we're already billionaires," he said. "We don't need the consulting money. We need to apply the Red Box anti-gravity device in a wide variety of transportation applications that will reduce the carbon footprint."

"Brother, you're right as usual," Matt replied. "We'll start on the briefing in the morning."

CHAPTER 2

MARS vs. MOON

A day later Matt called Reiner Strassburg, Chief Engineer at the Mercedes-Benz Research & Development Center in Sindelfingen, Germany, a suburb of Stuttgart.

"Reiner, my friend!" Matt began. "Now that your team and my brother have a possible solution for a Mars mission, please plan to accompany Murray and me when we brief the Director of NASA in Huntsville at the end of the month."

"Count me in," Reiner replied. "And if it's OK with you, I'd like to bring two associates: a propulsion engineer and a mechanical engineer who led the Turbopod redesign for space travel."

"That's a great idea. I'll call you soon when I have a meeting date and time," Matt said as he hung up the phone.

A week later, Matt got a call from NASA confirming the meeting at 10 a.m. on May 26 in Huntsville. He phoned Reiner to relay the information. "We're on at NASA on May 26," he said. "Please provide the names and vital statistics of your associates so we can clear them through security. And please plan to come here to Miami on the 24th so we can do a dry run of the briefing and confirm our strategy."

"Will do, Matt," Reiner replied. "I look forward to seeing both of you again and planning our next exciting joint venture."

On the morning of May 24, Reiner and his two colleagues, Dieter Wexler and Miriam Zimmerman, arrived at the Red Box manufacturing facility in south Miami. Murray greeted them in the lobby and escorted the team to the conference room where Matt sat at the end of the table. As they entered the room, Matt rose and welcomed his friend and new associates with open arms. Matt's wife, Heather, served coffee and bagels while the engineers set up their laptops.

Matt spoke first. "We've sure come a long way in the last two and half years—from the first Baldie seaplane trials to the Baldie barnstorming campaign to the Turbopod trials in Ordos, China, to the award ceremonies in Washington," he said. "Now we are asked to find a way to get to Mars and return safely. Reiner, Murray and I want to express our sincere gratitude for the work your team did in developing this Mars mission."

Reiner pointed to the device in the middle of the conference table and said, "All of our engineering and calculations revolved around your Red Box. Without it, it would be impractical, maybe impossible."

"Well said!" Matt exclaimed while powering up his laptop to begin the briefing. The first slide appeared as Murray walked to the front of the room to begin the briefing.

"First, let me thank Dieter for his support with all those propulsion calculations," Murray said as he clapped his appreciation. "Thank you, Dieter, and thank you, Miriam, for your T-Pod spacecraft designs. You did so many versions as our planning evolved. I particularly appreciate the genius of the simple mate-collar that enables multiple spacecraft to become a functioning cluster, or Mars Base."

Murray applauded again as everyone in the room looked at Miriam with nodding heads and a chorus of "Bravo!"

Murray then surprised his guests by saying, "We have prepared two briefings for NASA. The first one is the Mars mission, which we have developed over the past two months, and the second one..."

Just then Reiner interjected, "And the second one must be to the moon."

Murray's eyes grew wide before he furrowed his brow and asked, "How did you know?"

Reiner was ready with an answer. "The price tag and danger level of a Mars mission are too high given the return on investment. Our computers worked overtime to calculate the quarter-trillion-dollar estimate, which is low in my opinion and provides no redundant safety margins."

"Precisely," Murray said.

"And while we quietly support the European Space Agency's Mars mission planning, the real payoff for Earth can be a moon base," Reiner continued. "Our chairman agrees and is ready to support your moon mission, Matt."

Murray sat down, reflected for a few seconds, then said, "This is fantastic. But Reiner, I have to warn you about one thing."

"What's that?" the stoic German asked as he peered over his horn-rimmed glasses.

"Matt and I will only actively work on the moon mission. We'll formally submit both briefings to NASA, but we won't be involved in subsequent activities of any kind on the Mars mission. We will make it quite clear that our energies and resources must be focused on a successful moon mission because of its potential impact on Earth. Matt keeps reminding me about the dying polar bears!"

"I understand," Reiner replied.

Murray then added, "Of course, Mercedes-Benz is free to contract with NASA to support the Mars mission in any way you choose. It's your corporate decision alone."

"Thank you, Matt and Murray," Reiner said. "Let's see how this planning exercise plays out over the coming months. But let me make it quite clear that our company already has its hands full trying to meet all of the applications for the Red Box. And the more we develop electric vehicles, the more we help solve the global warming problem."

Matt then took the floor and brought up the moon mission slides to level-set the audience. For the next two hours Matt and Murray disclosed their detailed planning for a moon mission. It borrowed much from the Mars mission plan, but made more economic, social, and environmental sense.

The team was ready for the trip to Huntsville.

At 5 p.m. on the evening of May 25, two Turbopods, or T-Pods as they were known, took off from the Homestead Airport southwest of Miami and headed west over Everglades National Park. Captain Jim Russell piloted the T-Pod carrying Matt and Maggie; Murray's wife piloted the other. Reiner flew with Murray, while Dieter and Miriam flew with Matt. The two craft passed over Naples, Florida, at 1,000 feet and turned north over the Gulf of Mexico. Two hours later, the T-Pods landed at the Mercedes-Benz plant in Vance, Alabama. The T-Pods were secured in a large warehouse and the Red Boxes stored in a safe. The team had dinner as the guests of Production Manager Reinhold Timm. The team spent the night at a quaint bed & breakfast in Tuscaloosa 15 miles west of the plant.

At 8:00 AM the following morning, the two T-Pods headed northeast for the 30-minute flight to Huntsville. Reinhold was a passenger in Murray's craft. An hour before the scheduled meeting, the T-Pods were cleared to land on the lawn behind the headquarters. The team was escorted through security and set up several laptops on the director's conference table. Matt went to the window and saw the two T-Pods, and out of the corner of his eye he saw the two Red Boxes in Murray's valise.

Ten minutes later, Director Don Golden led his team into the room and warmly greeted the Flynn brothers and their colleagues. The parties took their seats at the conference table. The director began, "Three months ago we met here, Mr. Flynn, and asked you and your team to assist us with the planning for a Mars mission. "Pointing to his team, he continued, "We're eager to see what your solution is and how it might differ from our current baseline. Matt, the floor is yours."

"Thank you, Director Golden," Matt said as he motioned to Murray to start his briefing.

"Ladies and gentlemen," began the confident Matt, "we have a two-part briefing for you today, which I will explain in detail. I also will provide a written report within two weeks per our consulting contract."

For 40 minutes Murray outlined the Mars mission concept, which relied heavily on the use of Red Boxes and their ability to enable weightless launch vehicles, crew modules, and service modules. Highlights of the solution included:

- A redesigned T-Pod for long-term deep space travel with radiation shielding, small windows, and electric-powered ion thrust engines.
- One T-Pod to serve as the command module and the other to serve as the service module; both would sit atop the launch vehicle as it lifts off from Cape Kennedy.
- A launch vehicle with small engines would contain an imbedded logistics module.
- The three modules would travel as a caravan, using the moon to "slingshot" it into deep space.
- Halfway to Mars, the logistics module separates and station-keeps until the return of the command module. This module has oxygen resupply, water, food, fresh batteries, and a waste disposal system. The command module mates with the logistics module via a mate collar. The astronauts will rejuvenate there for two days during the return.

- Halfway through the 1,120-day mission, the command module controls the descent of the service module to the surface of Mars.

- The service module contains a laboratory, communications systems, power systems, backup oxygen, dust filters, inflatable solar panels, and an optical telescope which is left behind for subsequent missions.

- After confirming the readiness of the service module, the command module lands next to it and mates via the mate collar.

- The four-person crew (two males and two females) will spend 16 days on Mars conducting experiments, collecting samples, and following protocols sent from Mission Control in Houston.

- On Day 581 of the mission, the command module will fire its redundant aerozine thrusters to lift off the planet to rendezvous with the logistics module.

- On Day 841, the command module will separate from the service module and head to the Space Station, where the astronauts will be quarantined until a microbe analysis confirms their cleanliness. Some of the samples will be left at the Space Station for certain experiments.

- On Day 1,120, the command module slowly falls to Earth with the aid of parachutes and the throttle-down of the Red Box's antigravity capability. The module will land in water just off the coast of Cape Kennedy. At this point, the "Bald Eagle" module has landed.

At this point in the briefing, Murray asked the audience to go to the window and observe the two T-Pods joined by a mate collar put in place by engineers Timm and Zimmermann.

The director led a round of applause. As the group returned to the table, Murray took his seat and spoke in slow, distinct, and very measured sentences.

"Sir, our second mission plan is a revisit to the moon," Murray said. He then looked around the table to judge the reaction. There was only silence.

He continued, "The price of our Mars mission is based upon a few pages of assumptions. The price tag for this proposed mission is more than a quarter-trillion dollars. The mission is single-thread, and everyone in this room knows what that means. The mission is as dangerous as it is portrayed by the movie industry. For our team, the risk-reward ratio is unacceptable."

"Fascinating, Mr. Flynn. Please give us some more of your logic," Director Golden said.

"Well, we firmly believe there must be a reduction in the rate of global warming," Murray continued. "Our electric T-Pod–based transportation systems do their part. But as we see it, a Mars mission does nothing to solve the problem. Indeed, it draws many of our brightest minds away from the climate warming problem for years." He then went on to list the key research that could be conducted on the moon, including:

- Monitor the Earth's magnetosphere, which plays a critical role in keeping our atmosphere in place by deflecting solar winds.
- Observe lightning strikes, which occur about 4 million times daily and are an important element of climatology.
- Study ocean color change; phytoplankton are the beginning of the food chain for many animals, and their reddish clouds help track ocean currents.
- Develop new forms of hydroponic farming.
- Mine Helium 3; 1 ton of Helium 3 is worth more than $3 billion and could be used for future nuclear fusion power plants, which are 70 percent efficient (coal and gas are only 30 percent efficient).

Murray then asked Matt to make a closing statement. Matt rose from his chair directly across from the director and said, "Director Golden, we thought we would be doing the nation a disservice if we didn't present our moon mission concept. My brother and I, along with our German colleagues, strongly feel that our mission now is to help mankind survive on Earth. Everything that can be done on Mars can be done on the moon at a fraction of the cost. Let us remember, sir, that the original estimate for the International Space Station was $10 billion, but the price tag today is more than $100 billion and still rising!"

Matt then passed a document across the table. He said, "Here is our draft consulting contract for a moon mission. This agency can be assured of our very best efforts in this regard. In the package you will also find a commitment letter from our colleagues, providing both engineering talent and millions of euros to the endeavor."

Director Golden thumbed through the document and said, "We thank you very much for your plans. As usual, you have given us plenty to think about. We'll get back to you in the near future."

With no questions from the audience, Matt closed the meeting by wishing the agency good luck with whatever decision it made — Mars mission, moon mission, or perhaps both.

CHAPTER 3

MOON MISSION

Two weeks after the NASA briefing, Dr. Cliff Arnold, NASA's Chief Scientist, called Matt and Murray at their Miami offices. Murray joined Matt in his office to take the call.

"Good morning, Matt and Murray," Dr. Arnold said. "We've given considerable thought to your proposal and are keen to work with you to apply the Red Box to both our Mars mission and commercial moon mission.

"That's great news!" Matt replied. "We're eager to hear some of the details."

Dr. Arnold then outlined the proposal. "We will provide a consulting contract to your firm and subcontractors whom you deem necessary, such as your colleagues at Mercedes-Benz, to be re-negotiated annually. For the Mars mission we ask you to apply the Red Box to large launch vehicles such as the Atlas V and the Falcon Heavy Expendable Launch Vehicles, or ELVs. We anticipate purchasing 18 Red Boxes over the next year for tests. Our purchase order will allow for additional quantities to be negotiated annually. Is that OK with you?"

Matt replied, "Sounds good to us." He looked across the table to see Murray nodding his approval.

"Are there any plans to apply the Red Box to the crew capsules?" Matt asked.

"Yes," Dr. Arnold responded. "We've alerted a commercial space company, a current NASA contractor based in Hawthorne, California, that we would like to test the Red Boxes with its Dragon crew capsule. We would launch the capsule into low Earth orbit, then conduct a series of tests as the capsule circumnavigates the planet and splashes down near the Kennedy Space Center in Florida."

Murray then added, "Doctor, you may want to consider using two Red Boxes in each capsule to add redundancy. It's like wearing a belt and suspenders at the same time."

"You read by mind," Dr. Arnold quickly replied. "And there's one other area where the Red Box could play a vital role."

"Oh? What's that?" Matt asked.

"Per your Mars mission plan, we could use the Red Box in the logistics module, or LOGPOD, as we call it," Dr. Arnold answered.

"That makes sense to us," Matt said.

Dr. Arnold then added, "At the same time we are prepared to provide limited support for a commercial moon mission." He told the brothers NASA would provide:

- Mission Control expertise from the Johnson Space Center
- Lessons learned from past experiments
- Astronaut training in the Neutral Buoyancy Laboratory (NBL) swimming pool in Houston
- Launches coordinated with foreign countries

"Doctor, your support in these areas would be truly appreciated," Matt said. "And we look forward to your contract and will negotiate it in good faith."

Murray ended the conversation with an expression of gratitude and said, "Dr. Arnold, we planned to visit SpaceX at its new launch facility in Boca Chica, Texas, next week. Does that seem like a prudent direction to take?"

There was a pause on the other end of the line before the Chief Scientist said, "While I can't endorse one contractor over another, I can say that the company is innovative and forward-thinking as far as space ventures are concerned, and they have been a very responsive company with their Falcon rocket and Dragon spaceship. Good luck!"

Matt hit the button on the speakerphone and reached across the table to shake Murray's hand. Both men had broad smiles on their faces as they thought to themselves that the logical next application of the Red Box was about to be launched.

A week later, two T-Pods left Miami and crossed the vast expanse of the Gulf of Mexico. Captain Jim Russell piloted one with Matt, and Maggie Flynn piloted the other while her husband, Murray, periodically dozed off during the four-hour trip to the southern tip of Texas. The two crafts circled over the city of Brownsville at about 1,000 feet, than glided eastward, eventually touching down on Highway 4 just outside the gate at SpaceX's Boca Chica launch complex. Jim and Maggie powered down the T-Pods and secured the Red Boxes in carrying cases. They were waved through the front gate and escorted to the Control Center. As the party approached the front door, the Launch Director, Merritt Preston, greeted them, followed by a small team of engineers who clustered around the T-Pods in amazement.

"Welcome to Boca Chica, Mr. Flynn," Preston said as he shook hands with Matt. "It's a real honor to meet you. My management told me to roll out the red carpet for the inventors of the Red Box, and we'll do just that. You'll get the royal tour later, but for now please follow me to our conference room for sandwiches and discussions about the use of this miracle box for space missions."

The guests smiled and followed the director into a large conference room. They helped themselves to Texas barbecue sandwiches and sodas, and took their places at a table with two Red Boxes prominently placed in the middle. Matt began the discussion by saying, "We're here to make your projects more successful and profitable. You've read about the Red Box and seen videos of it in action. And there are two T-Pods in front of this building. Now is the time to collaborate to make space travel a viable commercial venture."

"We fully agree," Preston said. "And let me point out another important fact. A rocket is essentially a bomb trying to be contained. Your Red Box should greatly simplify the propulsion system and make the rocket much safer."

"Well said, Mr. Preston," Matt said. "At this point I would ask my brother to narrate a couple of videos about the Red Box so that our discussions can best proceed."

Preston nodded his approval, prompting Murray to approach the end of the table, where he used a laser pointer to highlight the video. The audience was, in a word, spellbound.

After 20 minutes, Murray ended his talk by saying, "The Red Box is the enabler that will allow people to escape from Earth. It eliminates huge engines and a lot of fuel so you can fly people and cargo profitably."

Matt then stood up, walked over to a Red Box, put his hand on it, and said, "Our T-Pod with its Red Box is ideal for travel here on Earth, but it won't survive in space. Your Dragon spaceship is ideally suited for our planned moon mission. Over the next day, I hope to convince you of that fact."

"You won't encounter much of an objection from this team, Mr. Flynn," the Director said. "We're ready to collaborate on so many levels. But please, let me ask a question: Why did you pick the moon and not Mars?"

Matt was ready with his now often-quoted answer: "Because the polar bears are dying. Anything that can be done on Mars can be done on the moon sooner, safer, and cheaper. And it can be more effective in saving Earth."

"Bravo!" a couple of engineers proclaimed.

Preston responded "I have the confidence of senior management to negotiate a mutually beneficial arrangement that will best use our $100 million commitment to Boca Chica."

"Well, for starters, my brother and I are prepared to match that figure, utilizing this site to the fullest," Matt said. "You'll see just how serious we are about having a successful moon mission and in the process helping to utilize this launch center for a long time."

The Director smiled as he surveyed his team and said, "Now it's my turn to say, 'Bravo!' What are some of the ways we can work together, Mr. Flynn?"

Matt replied, "Please, call me Matt. I have a detailed plan in these folders." Maggie passed the folders around the table. They contained more than 50 pages of detailed specifications and drawings, some of which showed what the Flynns envisioned for Boca Chica.

Matt then provided a list of actions:

- The Dragon spaceship will be redesigned to be constructed of composite tiles made by Mercedes-Benz in Vance, Alabama. The tiles must not only fit together perfectly, but they also have must have two internal layers with lithium ion for battery power and Kevlar for solar protection.
- The "Bald Eagle" spaceships, or "pods," will be assembled and tested in the SpaceX plant in Hawthorne, California. SpaceX

will design and build the engines, propellant tanks, battery packs, attitude control system, reaction control system, fuel cells, landing gear, and communications systems, all mounted on an aluminum chassis with titanium stiffeners. The Flynns will provide the software for autonomous operation and a small rover craft to be mounted on the side.

- Once the final design is established, the size of the engines and propellant loads will be established. The plan is for the Bald Eagles to dock at the International Space Station for checkout and fueling prior to proceeding to the moon. The engines will be tested at SpaceX's McArthur, Texas, facility and shipped to Boca Chica for integration and quality control.

- An optical telescope will be made in Germany and shipped to Boca Chica for shock testing and mounting in the Laboratory Pod (LABPOD).

- At the south pole of the moon, four pods will form a cluster moon base. The pods — two CREWPODS, one LABPOD, and one LOGPOD — will each have two access hatches for safety and flexibility. Each hatch will have a mate collar that forms a semi-rigid tunnel to another pod. These collars will be made in Vance, Alabama, and shipped to Boca Chica. Ultimately eight pods will be ordered.

- Four of the pods will compose a moon base mock-up located next to the Integration and Test Facility in Boca Chica. This mock-up will serve as a training facility for the astronauts.

- The Red Boxes will be manufactured in Miami and shipped here. They will be controlled with the strictest security practices possible.

- The parties involved will contract with NASA to provide Mission Control experts from Houston on an as-needed basis and to use its NBL to provide maneuver training for the astronauts.

- The craft will launch from Boca Chica. The jury is still out regarding whether the weightless Bald Eagles can be launched directly to the Space Station or whether a Falcon rocket will launch with two Bald Eagles as the payload. A team led by Murray will analyze the options.

"So there you have the short version of our mission plan, ladies and gentlemen," Matt said. "I'm sure you'll have thousands of questions. And in the end, it may be an enterprise in which you might not want to participate at all. I ask you to thoroughly read our plan under the aegis of a nondisclosure agreement and provide an answer by the end of the month."

"Well, Matt," Director Preston said, "the people at headquarters warned me about your persuasive salesmanship, but this is a bit overwhelming."

"I'll take that as a compliment," Matt laughed. "Actually, Murray and I would be glad to now break off into splinter groups to continue the discussions. We still have a couple of hours, if your schedule allows."

"That's a good idea," Preston replied.

"And if some of your folks are so inclined, Jim and Maggie would be glad to give T-Pod demo flights along the Rio Grande River," Matt offered.

A buzz of excitement filled the room. It was very clear that the guests were making a positive impression. Soon several engineers were huddled around Murray, asking questions — all adroitly handled by the genius.

Matt and Director Preston toured the facility in a Land Rover as they discussed the facility's growth plans.

Late in the afternoon, the two T-Pods lifted off of Highway 4 and headed northeast, eventually landing in Vance to brief Reinhard Timm and his managers while enjoying the hospitality of Mercedes-Benz.

At the end of the month, Matt got a call from the Director. "Matt, this is Merritt Preston," he said. "I'm delighted to report that our company will be a partner in your moon mission."

That's super news, Preston. You've made my day, perhaps even my year!" Matt replied.

The Director continued, "We also have done some preliminary calculations which I think you'll find rather exciting. How soon can you and Murray return to Boca?"

"We're commissioning a new Red Box manufacturing facility at a secret location early next week," Matt replied. "We could be there at the end of next week."

"Perfect," Preston answered.

"Great," Matt replied. "And I'll plan to have a propulsion expert with me, if that's OK."

"Sure. We want to get this project off the ground with right planning up front," the Director said.

After Matt hung up the phone, he called down the hall to Murray: "Hey, Brother! Project Polar Bear or as you call it, Project Bald Eagle II is a go!"

CHAPTER 4

POLAR BEAR XPRESS

It took two weeks for Murray's team to complete the statement of functional requirements and drawing package. After a final review with Matt and several aerospace engineers, the package was sent to the Polar Bear Xpress team in Hawthorne, California.

Three weeks later, Murray flew to California to answer questions and work with the team. Meanwhile, Captain Russell flew Matt via T-Pod JO-126 to Vance, Alabama, where they picked up the manager of 3-D production and a satchel of composite panels to demonstrate their form, fit, and function to the Polar Bear Xpress team. The T-Pod landed in the parking lot behind the fabrication facility a couple miles east of Los Angeles International Airport. Captain Russell carried the Red Box as the threesome entered the facility. They were met by Murray and escorted to the team room, where detailed technical discussions took place all afternoon.

That evening the Chairman of SpaceX, Alton Taylor, hosted a dinner for the team on the *Queen Mary* ocean liner at Long Beach. The T-Pod flew over Los Angeles' notorious traffic and landed on the water next to the majestic ship. Matt was convinced this would be the first in a series of banquets culminating with a moon landing celebration. As he left the party and boarded the T-Pod, he was struck by the progress of technology since the *Queen Mary* was christened.

The next day the T-Pod returned to Miami via Vance, where Matt was given a tour of the new 3-D production line reserved for the manufacture of space-qualified composite panels for the Bald Eagle spaceship.

The final design package was transmitted to Matt a week later. As he opened the email, he was thrilled to find that a Falcon rocket was not needed to boost a Bald Eagle up to the Space Station. Computer models confirmed that three aerozine engines would generate enough thrust for the trip. Matt then called Murray into his office. As Murray read the email, he smiled as if he knew all along that his invention wouldn't need the rocket's help.

"And to think, the Saturn V that took the astronauts to the moon required 12 million pounds of thrust and five stages to do the job!" Murray said.

"That's right," Matt added. "I wish Werner von Braun, chief designer of the Saturn V rocket, were alive to see the Bald Eagle on the moon."

Murray forwarded the email to his NASA colleagues for a final review prior to signing the prototype production contract.

Given the approval by NASA's engineering team, Matt organized the contract signing at the Boca Chica launch facility. Surrounded by local politicians and media, Matt and Preston officially signed the documents. And with cell phones and photojournalists still clicking pictures, Matt handed the keys to a T-Pod outfitted as an emergency vehicle to the mayor of Brownsville.

That evening the mayor reciprocated with a bull and oyster roast and a special bull-riding contest enjoyed by hundreds of locals, many of whom would be employed by the launch facility.

The next day, Matt and Murray took separate T-Pods for the half-hour flight to the Johnson Space Center in Houston to execute the consulting contract for Mission Control support and the use of the NBL for astronaut training.

Over the next 11 months Matt held monthly teleconferences with the team in Hawthorne developing the prototype Bald Eagle spaceship. A roll-out was scheduled for November 26. When the day arrived, it seemed like the entire world descended upon Hawthorne. In addition to the four Flynns — Matt, Murray, Heather, and Maggie — key personnel from Boca Chica and Vance were in attendance. Matt, who was paying all the bills and was now a national celebrity, served as master of ceremonies. The entire Space Company's Board of Directors took their places in the front row. Dignitaries included the NASA Administrator, the Governor of California, the FAA Administrator, and the Secretary of Commerce, a California native representing President Werner, who promised to attend the launch. Bald Eagle ONE (BE-ONE) sparkled in the morning sunlight as hundreds of Space Company employees filled the grandstand.

Matt praised all involved and called the event a global milestone, officially naming the mission the Polar Bear Xpress. As commemorative patches were passed out, Matt told the audience that after the initial test flights at Boca Chica, the BE-ONE would be gifted to the Smithsonian's Air and Space Museum.

Soon Matt relinquished the podium to Chairman Taylor, who spoke for a few minutes, recapping how much progress had been made in the past 11 months. He then motioned to have the hangar doors opened. Within seconds several employees pushed out a 7-by-4-foot wheeled platform, its contents shrouded in a gleaming Mylar blanket. The platform was rolled in front of the podium, where the chairman said, "Gentlemen, please remove the cover! In appreciation of our partnership with the Flynns, this trophy should also reside in the Smithsonian."

The blanket was pulled off, revealing a huge, taxidermied polar bear striding across a snowfield.

The audience roared its approval. Matt's eyes teared up with joy. Maggie hugged Murray.

The Chairman then closed the ceremony with these words: "Friends, this bear and Bald Eagle ONE will soon be on display as our moon mission takes a giant leap on behalf of planet Earth."

CHAPTER 5

ASTRONAUTS

For the Mercury program in the 1960s, only seven out of 500 candidate astronauts were chosen. They were a combination of pilots, engineers, and scientists. Before they were assigned to the Astronaut Office at the Johnson Space Center, they underwent a barrage of physical examinations and psychological screenings. Flight training and swimming, particularly military water survival and scuba certification, were essential prior to the rigorous two-year study of instruments, subsystems, robotics, and even virtual reality training to simulate spacewalks. Today a six-month mission on the International Space Station requires up to five years of training.

A year before the prototype Bald Eagle spaceships were delivered to the Integration & Test Facility at Boca Chica, the astronauts started their training for the moon mission. Out of hundreds of candidates, 20 were chosen for the first four crews. The first crew consisted of:

- Male, age 44, American, Commander
- Female, age 36, Russian, Engineer and Physician
- Male, age 39, German, 3-D Engineer on leave from Mercedes-Benz
- Female, age 30, Chinese, Environmental Scientist
- Male, age 40, American, Marine Corps T-Pod Riverine Pilot, Bald Eagle Pilot

The second crew consisted of:

- Male, age 34, Canadian, Payload Specialist
- Female, age 36, Brazilian, Pilot
- Male, age 42, British, Propulsion and Energy Engineer
- Female, age 31, Japanese, Robotics Specialist
- Male, age 44, American, Navy Seabee Construction Engineer

During the training year, the astronauts were housed in a small apartment complex on the south edge of JSC. Four T-Pods were stored in a special hangar at Ellington Field for flight training and shuttles between JSC, Boca Chica, and elsewhere. Four Bald Eagles were kept at Boca Chica as a simulated moon base, complete with mate collars joining the spaceships. And two Bald Eagle simulators were kept at the bottom of the NBL for astronaut training in a weightless environment.

During month three of the training, the pilots were sent to Miami to receive special training from Captain Russell and a tour of the Red Box manufacturing facility hosted by Murray. On the third day of this training, Matt and Heather hosted a swim party at their Key Biscayne home. The conversations centered on the use of the moon to research the Earth, but all eyes were on Brazilian pilot Claudia Santini, whose skimpy white bikini left almost nothing to the imagination.

During month four, the German 3-D engineer returned to the production facility in Vance to monitor the progress of the composite panels and give speeches about the moon mission to employees assembled in the cafeteria. Mercedes-Benz Chairman Werner von Boltz hosted a dinner in the astronauts' honor, and Matt and Heather attended.

After the dinner, conversation focused on the risks of space flight. The Chairman pelted Matt with questions on the subject. Heather listened attentively as Matt explained, "Mr. Chairman, our planet's physics is opposed to space flight. The physics of achieving orbit are very challenging to overcome. A huge rocket is really a dangerous bomb, and small deviations can result in catastrophic failure. But guess what?"

"What?" Chairman von Boltz asked.

"The Red Box changes the equation by eliminating the impact of gravity on the launch," Matt replied. "The Bald Eagles will slowly, albeit in a very controlled manner, ascend to orbit altitude with only small thrust engines."

"That's comforting," the Chairman said.

"But wait, sir. There's much more," Matt said. "In space projects, nothing ever works the first time you put it together. Again, the Red Box is a winner. Think about it. There are thousands of Red Boxes in use today around the world with a perfect operation and safety record."

"Good point," von Boltz said.

"And," Matt continued, "fear rules all decisions in space programs."

"What do mean?" inquired the Chairman.

"Well, sir, because the systems are so expensive and must operate in harsh environments, redundancy is used to limit uncertainty," Matt said. "This is particularly the case when new technologies are introduced where failure modes are unknown. Again the Red Box wins, and with a primary and secondary Red Box on each spaceship, the risk of failure is greatly reduced."

The chairman took a long draw on his Padron 1926 cigar as he looked Matt in the eye and said, "Thanks. I'll sleep better tonight."

CHAPTER 6

BALD EAGLE ONE

Right on schedule, the modified Dragon V2 spacecraft, now christened Bald Eagle ONE, finished its last set of qualification tests at the plant in Hawthorne. All of the seams of the composite panels had been X-rayed to confirm that their fit with adjoining panels met specifications. Flight software was uploaded to the on-board computers, and each subsystem was tested and retested as part of the pre-ship review.

As the paint on the spacecraft's transport container was drying in the heat chamber, Captain Russell was powering up a T-Pod to take Matt to California for the roll-out. They stopped in Vance to pick up the Mercedes-Benz production manager, Reinhard Timm. They then flew west southwest to Houston, where they received a flight readiness review briefing by Mission Operations at JSC. After an overnight stay near Hobby International Airport, the team, which now included two NASA Mission Specialists, continued the trip to Hawthorne. During this leg of the trip, Matt flew his craft along the south rim of the Grand Canyon at an altitude of 100 feet. Airplanes are not allowed to fly in or around this natural wonder; however, because the National Park Service had purchased a T-Pod to replace its expensive, noisy, dangerous rescue helicopter at the site, Matt was given special permission for the fly-over per an approved flight plan. Needless to say, the T-Pod's passengers were thrilled by the experience. The T-Pod finally landed in the parking lot of the production facility as the sun was setting over the Pacific Ocean.

The next morning, the five-person party arrived at the facility at 8 a.m. to witness BE-ONE be encapsulated by the transport carrier. A specification sheet was used during final inspection before a crane lifted the container up 12 feet as a flatbed truck rolled into the receiving bay. The container was gently lowered onto the truck. After tie-down and checkout of the container's environmental control system, Matt shook the driver's hand and the novel spacecraft began the 1,650-mile trip to Boca Chica.

A specification sheet was used during the final inspection prior to shipment.

BALD EAGLE SPACECRAFT SPECIFICATIONS

MISSION: Deliver crew and cargo to the ISS and the moon and safely return to Earth

SPACECRAFT: Composed of two modules: Crew and Propulsion/Energy. Crew module refurbished for future missions in 150 days; Propulsion module jettisoned during return and landing.

CREW: 5

MOTORS: 4 aerozine engines deliver 88,000 pounds of trust

POD COFIGURATIONS: CREWPOD, Laboratory (LABPOD) and Logistics (LOGPOD)

HEIGHT: 20 feet

DIAMETER: 12 feet

DRY MASS: 9,500 pounds

PAYLOAD: 7,300 pounds

SOLAR POWER: 2 deployable solar arrays generate 2Kw; jettisoned during return to Earth

VOLUMES: 350 cubic feet in pressurized CREWPOD; 1,200 cubic feet in unpressurized extend propulsion module

HATCHES: 1 side hatch, 36 inches wide; 1 nose entry hatch. Both have mate collars to join pods for cluster

LANDING STRUTS: 4 hydraulic deployable titanium; 2 feet long with stabilizer base

HEAT SHIELD: Phenolic impregnated carbon ablator

LAUNCH VEHICLE: None; redundant anti-gravity Red Boxes and periodic aerozine engine burns

COMMUNICATIONS: Multiple command uplink channels; fault-tolerant S-Band telemetry. Internal pod communications bus has standard RS-422 and 1553 serial input/output; Ethernet for internet-addressable payload services

FLIGHT COMPUTERS: 3 pair; radiation and fault-tolerant (up to 26 configured for missions)

REENTRY MOTORS: 4 super Drago to slow the decent while the Red Box is throttled down

Top View

Not to Scale

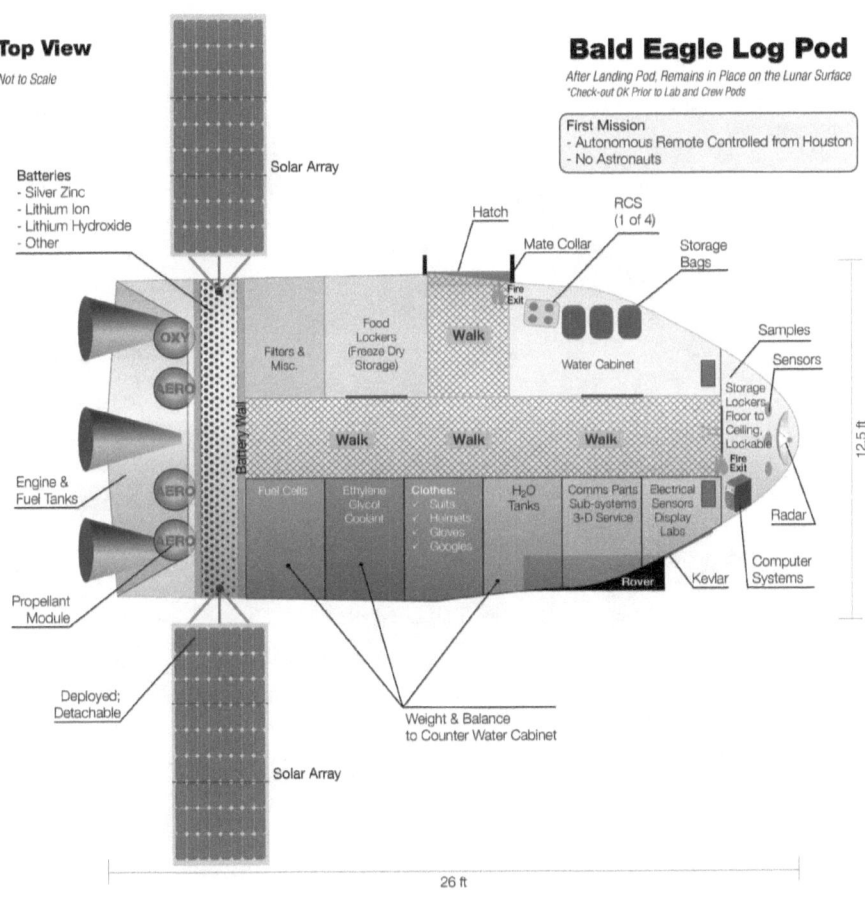

Bald Eagle Log Pod

After Landing Pod, Remains in Place on the Lunar Surface
*Check-out OK Prior to Lab and Crew Pods

First Mission
- Autonomous Remote Controlled from Houston
- No Astronauts

Batteries
- Silver Zinc
- Lithium Ion
- Lithium Hydroxide
- Other

Solar Array

Hatch

Mate Collar

RCS
(1 of 4)

Storage
Bags

Filters &
Misc.

Food
Lockers
(Freeze Dry
Storage)

Walk

Fire
Exit

Water Cabinet

Samples

Sensors

Storage
Lockers
Floor to
Ceiling,
Lockable

Fire
Exit

Walk

Walk

Walk

Engine &
Fuel Tanks

Fuel Cells

Ethylene
Glycol
Coolant

Clothes:
✓ Suits
✓ Helmets
✓ Gloves
✓ Googles

H₂O
Tanks

Comms Parts
Sub-systems
3-D Service

Electrical
Sensors
Display
Labs

Radar

Computer
Systems

Propellant
Module

Kevlar

Rover

12.5 ft

Deployed;
Detachable

Weight & Balance
to Counter Water Cabinet

Solar Array

26 ft

31

Side View
without Deployed
Solar Arrays

Bald Eagle Lab Pod

After Landing Pod Remains in Place on the Lunar Surface

Second Mission
- Autonomous Remote Controlled from Houston
- No Astronauts

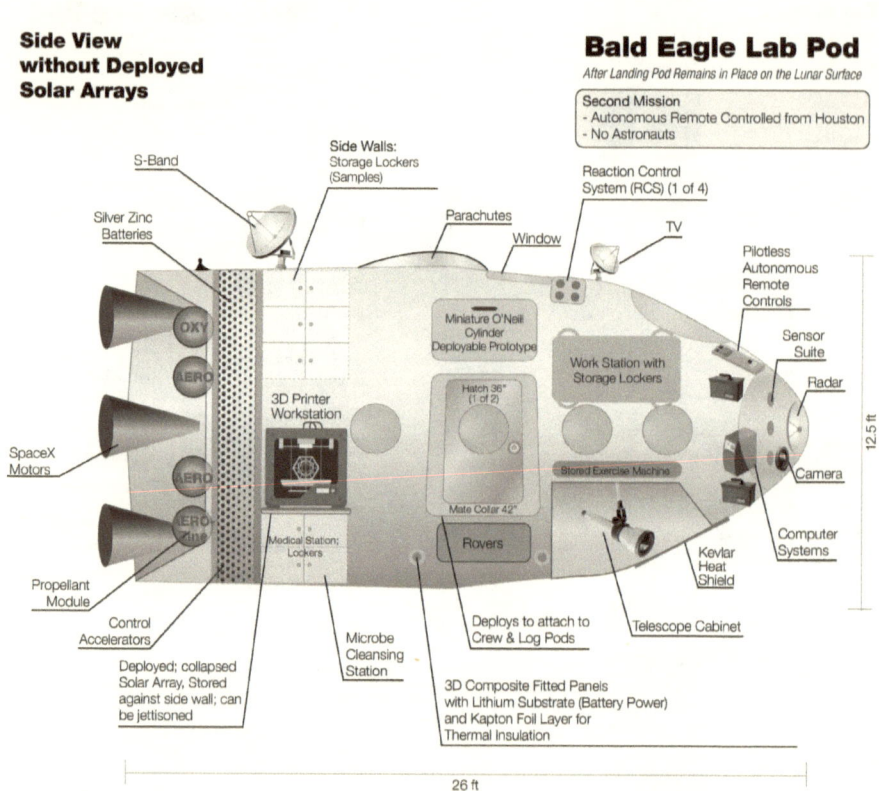

S-Band

Side Walls:
Storage Lockers
(Samples)

Silver Zinc
Batteries

Parachutes

Window

Reaction Control
System (RCS) (1 of 4)

TV

Pilotless
Autonomous
Remote
Controls

Sensor
Suite

OXY

AERO

Miniature O'Neill
Cylinder
Deployable Prototype

Work Station with
Storage Lockers

Radar

3D Printer
Workstation

Hatch 36"
(1 of 2)

12.5 ft

SpaceX
Motors

AERO

Stored Exercise Machine

Camera

AERO
Xtra

Mate Collar 42"

Computer
Systems

Medical Station;
Lockers

Rovers

Kevlar
Heat
Shield

Propellant
Module

Control
Accelerators

Microbe
Cleansing
Station

Deploys to attach to
Crew & Log Pods

Telescope Cabinet

Deployed; collapsed
Solar Array, Stored
against side wall; can
be jettisoned

3D Composite Fitted Panels
with Lithium Substrate (Battery Power)
and Kapton Foil Layer for
Thermal Insulation

26 ft

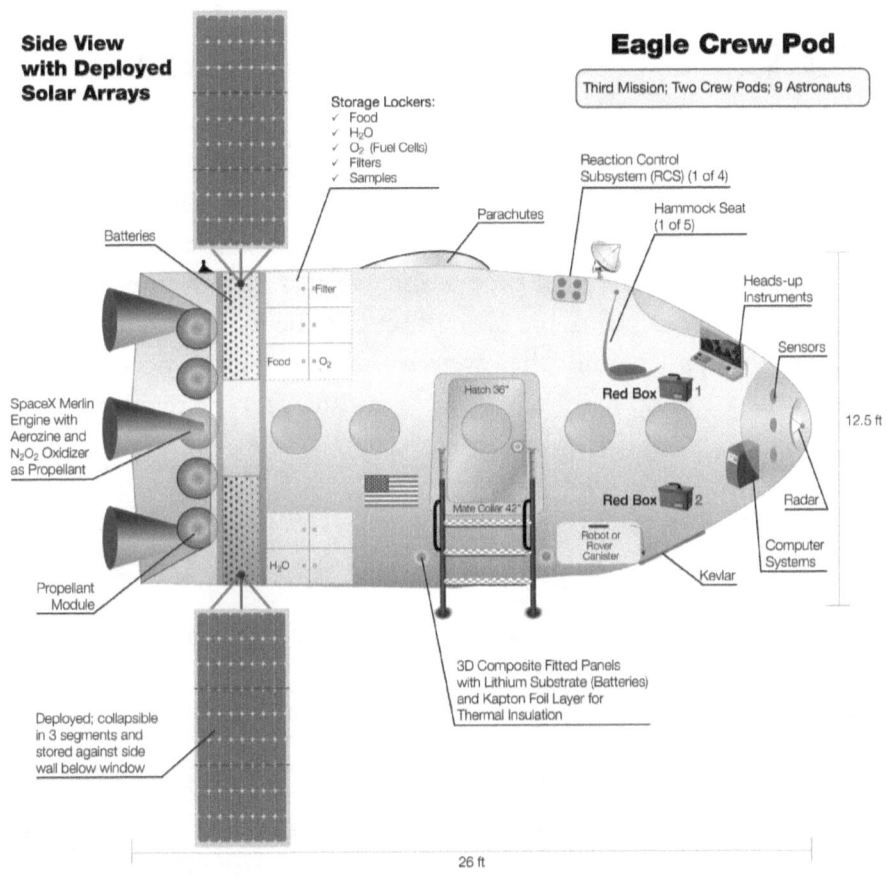

Side View with Deployed Solar Arrays

Eagle Crew Pod

Third Mission; Two Crew Pods; 9 Astronauts

Storage Lockers:
- ✓ Food
- ✓ H_2O
- ✓ O_2 (Fuel Cells)
- ✓ Filters
- ✓ Samples

Reaction Control Subsystem (RCS) (1 of 4)

Parachutes

Hammock Seat (1 of 5)

Heads-up Instruments

Batteries

Sensors

Filter

Food O_2

Hatch 36"

Red Box 1

12.5 ft

SpaceX Merlin Engine with Aerozine and N_2O_2 Oxidizer as Propellant

Mate Collar 42"

Red Box 2

Radar

Computer Systems

Propellant Module

H_2O

Robot or Rover Canister

Kevlar

Deployed; collapsible in 3 segments and stored against side wall below window

3D Composite Fitted Panels with Lithium Substrate (Batteries) and Kapton Foil Layer for Thermal Insulation

26 ft

33

CHAPTER 7

INTEGRATION & TEST FACILITY

BE-ONE and BE-TWO soon arrived at the cavernous, 100,000-square-foot Integration & Test Facility at Boca Chica. The spacecraft, which were manufactured to specifications and tested at the subsystem level in California, were mated to the propulsion module for system-level tests. Over a six-month period, they were tested in a wide variety of clean rooms and chambers to prove the craft's readiness for the rigors of space.

Early in the morning on a crisp, fall day, the I&T Manager, Bob Hock, greeted Matt and Murray with a bellowing Texan, "Howdy! You boys ready for the tour?"

"We sure are," Matt replied. Murray nodded in his usual quiet way.

"Great," said the rotund Texan. "You'll see just how careful we are with your precious Bald Eagles."

The three men went through an air wash tunnel and donned the smocks, booties, and hair coverings. Before they knew it, they were walking into a sparkling white room with huge lights suspended 30 feet from the ceiling. A large yellow crane ran across the top of the room with "40 Tons" lettered in black on the side.

Hock began, "Our first stop will be in Bay 1, where we mate the Bald Eagle to the propulsion module. Please step this way." As the party approached the bay, BE-ONE could be seen suspended from a crane over the propulsion module. It sat on a wind table platform which glided the heavy spacecraft to stations around the facility. Red streamers with the words "Remove Before Flight" hung at various locations on the propulsion module.

"You'll notice the locking pins on top of the propulsion module," Hock continued. "These are explosive bolts, which enable the module to be jettisoned during re-entry. And, of course, there is no fuel in the aerozine tanks. They will be filled at the launch pad."

"Mr. Hock, the BE-ONE hatch is 12 feet off the ground," Matt said. "Do I correctly assume that a gantry is used so the astronauts can enter the craft?"

"Yes, Mr. Flynn, that's exactly how it is done," Hock responded.

After the party walked around the module, the manager motioned the Flynns to follow him to the next station. Soon they were peering through a glass window into a large room whose walls were covered with 2- to 3-foot cone-like spikes made of a sponge-like material.

"This is our acoustic chamber where we test for deleterious impacts on the spacecraft with sound pressure approaching 150 decibels," Hock said. "As I'm sure you know, Mr. Flynn, this is typically the noise from a launch vehicle. In the case of the Bald Eagle, however, the aerozine thrust will be minimal and very quiet compared to a launch vehicle. Your Red Box makes this possible."

"Thank you, sir," Murray replied.

The next station was also an enclosed room with a support structure in the middle. "This is the Vibration Table, which shakes the spacecraft to simulate launch impacts." Hock explained. "Again, because of the Red Box, vibrations during lift-off will be minimal. I can't wait to see it happen."

"You and me both," Matt said.

"Now please follow me, gentlemen," Hock said. About two minutes later the trio was standing next to a thermal vacuum chamber. The round, vault-shaped, metal structure was 30 feet tall with a 15-foot door on one end.

"Matt, Murray, this is where we'll subject each Bald Eagle to extreme temperatures in a perfect vacuum, just like out in space," the manager said.

"How extreme?" Murray asked.

"The temperature range goes from 100 degrees Celsius to minus 150 degrees Celsius," Hock replied. "An astronaut would not survive without a spacesuit at those temperatures. We'll do six cycles over an eight-day period."

The manager continued, "There are many other rooms in this facility — instrument calibration laboratories, subsystem testing laboratories, computer systems rooms for software integration and testing. However, our tour today only has enough time for the station directly across this bay."

As the three men approached another large glass window, Hock said, "This is the electromagnetic interference test chamber, or EMI chamber. Your Bald Eagles can't have any disruption in operation due to extraneous electro impulses impacting mechanical systems, much less all of the on-board flight computers."

After a brief question-and-answer session, the team re-entered the garment room, removed their smocks, booties, and caps, then walked to the visitor's lounge.

Matt thanked their new colleague, saying, "Thank you and goodbye, Mr. Hock. It's clear that our Bald Eagles will be ready and eager to fly when they leave this facility for the launch pad."

CHAPTER 8

RENDEZVOUS

For several late nights in the virtual reality laboratory at JSC, astronauts Hans and Katia worked on an Extra Vehicular Activity (EVA) at the moon base to deploy a telescope from the LABPOD. The telescope was part of a series of experiments to verify Einstein's General Theory of Relativity that light actually bends due to gravity as opposed to always traveling in a straight line. Often their hands, arms, and torso would touch in order to align the telescope and/or adjust the fine focus. A kind smile would come over their faces.

"You know, Katia, you're pretty good with these fine adjustments while wearing these clunky gloves," Hans said.

"Thanks," Katia replied. "I think it comes from being a farm girl in the Ukraine, where had to keep a lot of the machinery operating. You're not so bad yourself."

"Thanks," Hans said. "I'm a good marksman with a rifle and a bow and arrow. I think a steady hand helps in this task."

"It sure does," Katia agreed.

"I think we've done all we can do until the next task is assigned," Hans said. "I'm starved. Do you want to join me for pizza at Pomodoro's? It's close on NASA Parkway, and the food's great."

Katia thought for a second, then replied, "Well, I did skip dinner today. Sure. Sounds like a great idea. I'll follow you to the restaurant. Oh, and were going Deutsch — I mean Dutch — on this, right?"

"OK, we're Dutch!" Hans laughed.

Twenty minutes later, the two astronauts sat in a corner booth talking about everything except work.

Hans began, "Katia, rumor has it that you're an outstanding cosmonaut. Where did you receive your training?"

"Well, I spent a year in a very intense program at the Gregarin Cosmonaut Training Center in Star City, a former secret military base in Gorodok. That's just outside Moscow," Katia said. "The facility even has a full-size Space Station in a zero-gravity simulation tank. I feel like I've already been in space and have the right training to help the moon mission."

"No doubt," Hans said.

"Aren't you excited about the prospect, Hans?" Katia asked.

"Excited? Hell, I feel honored," Hans exclaimed. "I've been watching my company's activities with the Red Boxes and the T-Pods for almost two years. And our composite panels form the Bald Eagles. Every morning in front of the mirror, I thank heaven for being here. And, if I may say, working with you and the other pod teams has been a real joy."

Katia was quick to answer. "I feel the same way," she said, brushing her long blond hair away from her face.

After enjoying a pizza and a few soft drinks, the couple looked down at their watches to realize the late hour and the full schedule that awaited them the next day. Hans walked Katia to her car and made sure she was safely inside. She rolled down her window to say goodbye. As they parted, Hans said, "Katia, this was really fun. Can we do it again the next chance we get?"

"Sure. I'd enjoy that very much," Katia said with a smile as she closed the window of her Porsche and drove away.

Three weeks later, Hans and Katia worked together in the NBL on the O'Neill cylinder, one of the most important experiments planned for the moon base. In 1954, a German scientist, Hermann Oberth, described the use of a gigantic habitable cylinder for space travel. In 1976, American physicist Gerard O'Neill adopted the two counter-rotating cylinders in studies at Princeton University. The cylinders rotate in opposite directions in order to cancel out any gyroscopic effects that would otherwise make it difficult to keep the O'Neill cylinder aimed at the sun. The rotation provides artificial gravity on their inner surfaces. Habitats and manufacturing would be located in the middle, while the outer surface would support farming. The habitat would have to rotate 40 times an hour to simulate standard Earth gravity.

While O'Neill's cylinders were planned to be 5 miles in diameter and 20 miles long, the two astronauts were implanting a 10-foot long cylinder to test it effectiveness in growing plants. Prepackaged with seeds and water droplets, the cylinders were collapsible within themselves to only 3 feet for transport on the LABPOD.

"There. I think I got this end tethered," Katia said as she watched her partner struggle with the other end as mild turbulence was introduced by a scuba diver in the NBL to simulate solar wind. Katia joined Hans, and the two of them were able to secure the tie. A few seconds later, the diver gave the thumbs-up sign to the astronauts, indicating they were successful in this timed activity.

The next activity required the pair to attach an oxygen cylinder to a port orifice on the O'Neill cylinder. This was a simple but important task. Hans was then handed a nitrogen cylinder for attachment opposite the oxygen cylinder. Hans watched as Katia attached it. On the moon, these cylinders would be essential to simulate Earth's air pressure at sea level.

And this atmospheric mixture would provide shielding from harmful doses of cosmic rays. The final activity entailed attaching two 18-inch mirrors to the cylinders to reflect sunlight inside. In O'Neill's concept, night is simulated by opening the mirrors, letting the inhabitants view empty space. With the activity complete and performed ahead of schedule, the astronauts floated to the surface and were raised out of the NBL via a swim platform. The two astronauts raised their visors, looked each other in the eye, and smiled at their joint accomplishment.

Bald Eagle astronauts in training were required to participate in four hours of flight time per month in a T-Pod. This applied to payload specialists was well as pilots. The special fleet of T-Pods was housed in a hangar at Ellington Field adjacent to JSC. All astronauts qualified for solo flights after eight hours of flight time with Captain Russell or one of his pilots. A prescribed course had the T-Pods lift off the runway at Ellington Field, rise to 100 feet, turn south, and fly over a remote part of JSC. They then flew out over Clear Lake, Trinity Bay, Galveston Bay, and then the Gulf of Mexico. Four two-hour sessions were flown, usually in a series of figure-eights to allow the pilot to train in a wide variety of circumstances. The T-Pods usually returned over Goat Island and its multiple national wildlife refuges.

It was midpoint in the astronaut training program on a sunny Sunday morning in humid August when Katia's phone rang.

"Hello?" she answered.

"Hello, Katia," a male voice replied. "This is Hans. I was wondering if you wanted to get some flying time in and join me for a late brunch. I've found a secluded cove on Goat Island that really looks neat."

"Boy, do I need a break," Katia responded. "It sounds lovely."

"Great," Hans said. "How about I pick you up in an hour, and we'll go to Ellington?"

"OK," Katia replied.

"Oh, and bring a bathing suit. The water is clear and cool," Hans said as he finished the conversation.

An hour and a half later, the pair checked out a T-Pod, inspected the weather forecast, and filed a flight plan. Hans flipped a coin, and Katia won the toss. She would fly out, but Hans would fly back.

The sky was muggy and overcast, but no other planes were in sight as Katia flew over the coastal cities and out over the Gulf of Mexico. In a log, Hans witnessed the flight times and maneuvers. A half-hour into the flight, Katia cut off one of engines, per procedure, and started a smooth glide path for two minutes before restarting the number 1 engine. Her piloting skills were noted in the log.

At the hour mark, Hans directed Katia as the T-Pod neared the cove where they would put down. The craft circled at 100 feet as Hans pointed to a clear space for anchorage, ideal for the T-Pod, which was also a hydroplane. In three minutes, T-Pod 007 was floating as Hans dropped the anchor off the swim platform at the rear of the craft. Katia texted the coordinates to NASA's Flight Operations.

Their green flight suits were quickly removed, revealing bathing suits. Hans was prepared with swim fins and snorkels for both of them. As they sat on the swim platform, coating each other with sun tan oil, Katia looked at the fins and asked, "How did you know my shoe size?"

"Easy," Hans explained. "I looked in your spacesuit boot."

They shared devilish smiles as they dropped into the water and swam toward a colorful reef. They would occasionally point to something, gulp a breath of air, and swim to explore the site. At one point Hans saw something glimmer in the sand. He snatched it up and brought it to the surface. With genuine excitement, he showed Katia what he believed was an old coin, perhaps Spanish, perhaps gold.

He knew it was a lucky sign.

After 20 minutes they swam back to the T-Pod. Katia climbed up the ladder first. Her tight, slender body, covered only by a striped orange and white bikini, mesmerized the slight, pale German. As he climbed the ladder, he thought to himself, "What a goddess. And she's smart too!"

Soon the picnic basket was opened, and Hans showed his culinary skills with a wide variety of hors d'oeuvres, cold cuts, and pastries. He was German, after all! He even brought real glasses to serve the Perrier. The layout of the T-Pod enabled a 6-foot-square towel to be laid out for the picnic, allowing the pair to lounge in comfort, using the seat cushions for support.

As they clinked glasses, Hans toasted, "Here's to a successful moon mission and adventures in space, wherever we go in the universe!"

They both laughed.

In a moment of sincere warmth, Hans reached over and caressed Katia's long, wet golden hair with a tenderness rarely felt by the AAA Russian woman. She felt warmth cover her entire body, and she tingled with excitement. Hans gazed into her eyes as his hand slid down from her neck to her shoulder. He motioned for Katia to turn around. She obliged, and Hans used his large, strong hands to massage her neck and back, even under the bikini strap. Katia clearly enjoyed the experience, which elicited an occasional groan of delight.

"Oh, I didn't know I had a muscle there!" she exclaimed.

Hans then gently turned her around and kissed her on the forehead, saying: "Katia, you're really special, and I think you know I'm really fond of you."

"I know," Katia answered in a soft voice.

Hans then kissed her with all the tenderness he could muster as they slowly managed to lower themselves onto a pillow in a warm embrace, with their partially clad bodies touching in many locations. He caressed her breasts and could feel her nipples harden. Their kisses became more intense as Katia reached down and stroked his groin, saying in a soft, sweet voice, "Hans, let's not go too far today. We'll have other sweet moments in the future."

The German smiled, slowly lowered his head in consent, and said, "You're right. And I look forward to those moments very much."

At 4 PM, Hans landed the T-Pod on the Ellington tarmac right on schedule as thunder claps were heard in the distance.

CHAPTER 9

LAUNCH

Six months after it arrived at the I&T Facility, BE-ONE rolled out, right on schedule. The event was Webcasted to the plant in Hawthorne, the plant in Vance, and Mission Control in Houston. The media praised the new commercial space industry as well as the need for a moon base to help solve the problem of global warming. The Flynns had their own team of videographers to capture the occasion.

A low-ride transporter delivered the spacecraft to the launch pad. Scaffolding was wheeled into place to encapsulate the craft. The countdown clock was started at T-26 hours, a short period due to the absence of a towering launch vehicle.

At T-26, the following actions took place:
- Backup flight systems checked
- Flight software and memory unit functions reviewed
- Navigation systems activated and tested
- Preliminary inspections of flight deck completed

At T-20:
- Launch pad cleared of all nonessential personnel
- Launch pad swept of any debris

At T-16:
- Built-in hold in preparation for propellant loading

At T-12:
- Aerozine loaded in the three propellant module engines
- Ground support equipment closed out

At T-10:
- Weather and engineering briefings held
- Inertial measurement preflight calibration performed
- Boca Chica tracking antennas aligned

At T-6:
- Detailed analysis of Bald Eagle, including cockpit switch configurations, conducted by final inspection team

At T-4:
- The Flynns arrived in separate T-Pods and proceeded to the Control Center for briefings on the mission plan and the spacecraft status

At T-3:
- Televised weather briefing held
- Communications systems checks completed
- Voice checks with the Control Center in Boca Chica and Mission Control in Houston performed
- Secret Service sweep of the launch site completed
- Planned hold

At T-2:
- President Werner and staff arrived
- Secretary of Defense arrived
- Chairman von Boltz of Mercedes-Benz arrived
- NASA's Chief Scientist, Dr. Cliff Arnold, arrived
- Congressional delegations, including the senators from Texas and California, arrived

At T-1.5 hours:
- Astronauts entered the spacecraft
- Crew hatches closed and checked for leaks

- Scaffolding rolled back and secured in the fallback area
- Television broadcasting began

At T-1:
- Highway 4 shut down by Texas State Police
- Speeches given by President Werner and Texas Senator Jim Bob Raeburn

At T-30 minutes:
- Launch Director Preston conducted final launch team briefings
- Inertial measurement unit alignments completed
- Bald Eagle's onboard computers transitioned to launch configuration
- Backup systems transitioned to launch configuration

At T-9 minutes and holding:
- Final launch window determination and confirmation completed
- Flight recorders activated
- Launch Director given "go/no go" from the launch team
- Mission Control in Houston confirmed readiness: "Bald Eagle, you are go for launch."
- Automated ground launch sequence started
- Range safety devices armed
- Backup power units started

At T-50 seconds:
- Bald Eagle transferred from ground to internal power
- Crew members closed and locked their visors
- Auto-launch sequencing started

At T-10 seconds:
- Pilot activated the Red Box and affirmed local control

T-Zero: LIFT-OFF

Shouts of joy could be head from the spectator grandstands a mile away from the launch pad. The ascent was perfect and uneventful. The

Assistant Launch Director handed out ceremonial cigars. BE-ONE could be seen through the Control Center's windows gleaming in the morning sunshine as she rose to the 2-mile threshold.

President Werner's voice came over the loudspeaker, congratulating the participants and wishing the astronauts and the mission Godspeed. He ended by saying, "If all goes well, we hope to have the moon base operation in less than two years."

The dignitaries departed shortly after noon. The Flynns and von Boltz co-hosted a banquet that evening at Vic & Antony's Steakhouse in Houston. The Mayor and the Fire Marshall were in attendance, legitimizing the smoke coming from a sea of cigars.

As Chairman von Boltz left the restaurant, he shook Matt's hand and patted Murray on the back as he said, "Move over, Wright brothers. The Flynn brothers have just started another era in the history of flight!"

CHAPTER 10

ON ORBIT

With a successful launch two astronauts were on their way to the International Space Station and into the history books. Upon reaching the desired Low Earth Orbit (LEO) at 188 miles, the Commander radioed Houston, "I know that this was originally planned as a robotic mission, but I'm sure glad I'm here."

Mission Control at JSC came back, "We are too. All systems are nominal. You are good to go for two orbits prior to ascending the remaining 60 miles to the ISS."

"Roger, I copy and will execute," the Commander, Tom "Catfish" Crowley, replied as he looked over to his pilot, Herman "Fish" Salmon. They had been astronaut buddies for almost a decade, and this was their five days of fame.

"Bald Eagle, this is Houston," came over the Commander's earphones.

"This is Bald Eagle," the Commander answered.

"Bald Eagle, your orbit coordinates have just been uploaded and confirmed. Do you copy?"

"Roger, Houston. We have the flight plan and are go for ascent," Crowley said

"Godspeed," came over the earphones.

At this point the spacecraft was behaving just as the models had predicted. It was stable, with all systems performing as designed. As the astronauts looked out of the windows and saw New York City lit up at night, Crowley told his pilot, "I've been there many times!"

The pilot responded, "Me too. I love New York and so many other places in this great nation of ours."

BE-ONE completed its first orbit around the Earth with all systems nominal. Two hours later the aerozine engines fired to put the craft on an intercept vector with the ISS as it circled the globe at an altitude of 248 miles at an inclination of 51.6 degrees.

Eighteen minutes later Crowley reported, "Houston, we have 'mother' in sight and are approaching per flight plan."

"BE-ONE, we copy. You are go for mate at port one. Do you copy? Over," Houston announced.

"That's most affirmative on your last," Crowley replied.

Fifteen minutes later, BE-ONE was securely docked at the $150 billion ISS, launched in 1998 and expanded ever since into a behemoth that can frequently be seen from Earth.

"Welcome aboard, BE-ONE. We're glad you're here!" came over the two astronauts' earphones.

"So are we!" replied the skipper.

"Do you have the goodies?" the voice from the ISS inquired.

"We do," Crowley said.

Fifteen minutes later, the hatch was opened by Alexie, a Russian cosmonaut. The three men then formed a chain to unload the supply canisters from the Bald Eagle. Nothing's easy in space. The effort took two and a half hours, after which Houston called and said, "Your vitals are fine. It's time to rest. Tomorrow will be a busy day."

"Roger that, Houston," Crowley replied. "We're going to eat dinner and bed down."

Salmon performed the requisite shut-down procedures, and sleep came easily for the pair.

On day two on the ISS, the "Fish Twins" — Catfish and Salmon, as they were now nicknamed by the Boca Control Center — were busy unloading the rest of the supplies. In the afternoon they accepted waste canisters and secured them in the designated lockers. The two BE-ONE crewmen then enjoyed a tour of the ISS. As part of the mission, Catfish used a laser to measure the exact dimensions of the port area so that engineers would have parameters for sizing the "garage" envisioned to hold four Bald Eagles at one time.

At 18:00 hours, BE-ONE prepared to separate from the ISS and begin its deorbit maneuvers to eventually land at Boca Chica. While in orbit and attached to the ISS, BE-ONE traveled more than 15,000 miles per hour. For proper descent, it would have to slow to about 500 miles per hour on a computer-controlled trajectory. If the angle was too shallow, the craft could skip off the atmosphere and back out into space, perhaps never to be seen again. And, of course, the ceramic heat shield on the bottom had to remain intact. Even a small hole could lead to real problems. Salmon had done this maneuver a hundred times in the simulator, but now it was for real.

After one orbit around the Earth at a 30-degree inclination, the Red Box, now under computer-control, enabled gravity to slowly pull the Bald Eagle back down into the atmosphere. Off the west coast of Africa,

Houston confirmed the spacecraft's vital statistics and informed the crew that they were "go" for touchdown. Within minutes the Bald Eagle flew over Cape Kennedy, where tracking cameras confirmed the craft's tiles where intact. As it passed over Tampa, the four super Drago thrusters ignited to slow her down and position her over Boca Chica. At 250 feet, the four landing gears were extended and locked in place. It was noon in Texas, and a cloudy sky provided a nice backdrop as the proud bird dropped down in a perfectly straight line onto the concrete launch pad. Thousands of spectators in the grandstand cheered the event.

Only the launch team was allowed to approach the spacecraft with fire control equipment and the scaffolding. The crew shut down all the systems, and after eight minutes emerged from the side hatch. They stepped on the scaffolding with the aid of contractor personnel. They waved to the crowd, and cheers could be heard in the Control Center. The Fish Twins walked around BE-ONE to inspect it for damage. There was none. BE-ONE would eventually fly again on a future mission, perhaps to the moon. Computers in the Control Center confirmed that a residual of 10 percent fuel remained. The astronauts, each carrying a Red Box, took their seats on an extended golf cart–type vehicle, which took them to the Integration & Test Facility.

All the astronauts came from Houston to witness the event, and celebrations took place all around America. Murray and Maggie were in the VIP Room at JSC. Heather served up a red cake shaped like the Red Box at the production facility in Miami. Hawthorne was crowded with key spectators, including the Secretary of Defense and his general staff. Matt and Captain Russell were at Boca, where the obligatory barbecue was held that evening.

At the barbecue Matt got a call from his friend, Chin Chin Po, the diminutive detective in Hong Kong who saw the landing on television.

"Matt, that was outstanding. You Flynn brothers are so amazing," Po said. Matt thought for a second and then said something that took

Po's breath away. "How about you becoming an astronaut?" he asked. "You're only 31, have a PhD, and you're in perfect health. Based upon your work on behalf of the Red Box, the President could make it happen. Chin Chin, I'm not going to give you time to think it over. It's either now or never. There are many reasons why we need you as an astronaut. May I meet you in Houston?"

Over Matt's cell phone came the one simple expression that might change history: "OK."

CHAPTER 11

APARTMENT 6A

On a quiet Saturday evening two months after the BE-ONE landing, during the eighth month of the Bald Eagle astronaut training program, a soft knock on the door of Apartment 6A in the astronaut complex brought the British propulsion expert, Lennon Blair, to his feet. He walked over to the door, peered through the peephole, and saw the top of a head with black hair. He slowly opened the door.

"Miko, what a pleasant surprise. How can I help you?" Lennon asked.

"You already have, Lennon," she said. "I just wanted to thank you for your help in the math lab last week. I wouldn't have passed the exam without your help."

"You're most kind," the Brit responded. "Please come in."

The diminutive Japanese robotics specialist entered the apartment carrying a cloth satchel. She turned to Lennon and handed the satchel to him as she said, "Please accept this as a token of my appreciation."

The British engineer motioned for his Japanese colleague to have a seat on the sofa. He sat next to her and carefully opened the gift. He pulled out a round bottle of sake and two porcelain cups.

"This is really neat, Miko. I've never had sake," Lennon said.

JASON O'NEIL

"I think you'll like it, Lennon. A couple of warm cups will really relax you," Miko said.

Lennon pulled out the cork and sniffed the top.

"It has a crisp aroma. Almost like rice," Lennon said. He then asked, "How do you serve it?"

"You can serve it hot or cold," Miko replied. "I prefer it warm — in Japanese, *nuru kan* — just above body temperature at 104 degrees. We can warm the bottle in a pan of hot water."

"Do you have time now?" Lennon asked. "Shall we try it now?"

"Sure, I was hoping you would," Miko said with a slight, almost devilish smile on her face.

Ten minutes later the couple left the kitchen and sat down on the sofa. They clinked the little cups, and Miko put the cup to her mouth and swallowed the contents. Lennon followed her lead.

"Wow!" he said as an immediate warmth flushed his face. "This is really good. I had no idea. What a great thank-you!"

"I'm glad you like it," the tiny robotics expert said. "May I be so bold as to tempt you with another cup?"

"Well it's Saturday night, and I've got all tomorrow to study my astrophysics," Lennon said. "Sure, it would be a pleasure — indeed, an honor."

The second cup was followed by a third before the bottle was put in the cupboard.

Both the host and his surprise visitor were feeling relaxed, almost carefree, when Miko asked, "Are you ready for the second half of my thank-you, Lennon?"

The Brit was truly intrigued. "But what could it be, Miko? You've already been too generous."

"Have you ever heard of Ashiatsu barefoot back massage?" she asked.

"No, I haven't," Lennon said. "You're full of surprises aren't you?"

"Well, Lennon, as you know, astronaut training can be quite stressful. The sake calms you, but a barefoot massage *really* relaxes you," Miko said.

"OK, I'm game," Lennon replied. "What do we have to do?"

At that point Miko took Lennon's hand and led him over to the apartment wall. She put a small pillow on the carpet.

"Lennon, please take off your shirt and lie face-down with the pillow under your tummy," Miko instructed as she took off her shoes. With the wall to sturdy her movement, she stepped up onto Lennon's upper back, careful to avoid the spine. She then used her 90 pounds and strong toes to massage the muscles down his back. She slowly turned around on Lennon's buttocks and traversed back up to his shoulder.

"Incredible," Lennon moaned as he consciously felt stress leave his body. Indeed, in an instant the rocket expert forgot about the propellant mixtures that typically occupied his mind.

"There. How do you feel now?" Miko inquired as she bent over to pick up her shoes.

"Let me show you my appreciation," Lennon answered as he gently tugged on Miko's blouse, beckoning her to join him on the floor. In an instant he was running his hand through her long, black, silky hair and kissing the side of her face. Miko turned her head toward Lennon, and their eyes and lips simultaneously met.

"This is really happening," Lennon thought. "It must be a dream. It's too good to be true." He had always admired the tiny, classy woman from a distance. And now they were in a sweet embrace.

In a quiet voice, Lennon told Miko, "I can see I'll have to give you more math tips, because these gifts are, like you, incredible."

Miko blushed slightly and leaned forward to kiss her new partner. She placed Lennon's left hand on her breast and said, "There, now *you* can massage *me*."

The impact was immediate, as Lennon clutched Miko with an embrace he never wanted to end.

But it did. A few moments later Lennon got up, walked over to the apartment door, and locked the deadbolt. He then took Miko by the hand and led her into the bedroom.

The next morning Miko rose early, got dressed, kissed her new lover, smiled sweetly, and left Apartment 6A.

They each knew something very special had just begun.

CHAPTER 12

TSUNAMI

On a quiet Saturday morning in August, two engineers from Norway were enjoying breakfast on the deck of the Schooner Café in the small port city of Saudarkrokur, Iceland. The engineers, Tryg Ager and Nils Norgard, were conducting a site survey to expand the harbor of this idyllic town about 110 miles north of Reykjavik.

While warming his two hands on a cup of coffee, Tryg asked Nils, "Do you see that?"

"What?" Nils asked.

Tryg pointed to the edge of the breakwater, where the mast of a sunken sailboat was clearly visible. "There, about a hundred meters down the breakwater — it's the mainsail mast of a sailboat."

"Oh, now I see it," Nils said.

"Something's wrong," Tryg said. "It's not low tide, but the water is receding out of the harbor. It's simply too low for this time of day. And, there, look over there! Ralf's skipjack — it has a draft of 6.5 feet — is leaning up against the pier."

A worried look came over Tryg's face as he took another sip of coffee. He listened for birds chirping, but heard nothing.

An hour earlier, a massive chunk of Greenland, an icefield about the size of the state of Connecticut, broke off the country's east coast and plunged into the North Atlantic. The result was a wave 50 feet high, traveling southeast at almost 150 miles per hour, headed directly at Iceland. Saudarkrokur was the bull's-eye.

"Nils, look," Tryg said, pointing toward the water. "There are whitecaps at the half-mile buoy."

"Oh, shit. Let's get the hell out of here!" Nils yelled. The two threw their money on the table and ran across the deck. As they passed the café's front door, Tryg hollered, "Tsunami! Run for your lives!"

The nautical engineers ran the half a block to their Range Rover and immediately started it up. They made a U-turn and sped eastward out of town. As Nils pushed the gas pedal to the floor, Tryg called the emergency telephone number on his cell phone.

Only a minute later they were climbing out of the valley. Tryg estimated they were about 100 feet above sea level when he asked Nils to pull over. He leaned out of the passenger side window and put his cell phone in camera mode. His jaw dropped and he could barely speak.

"Nils, it's gone!" Tryg said. "Saudarkrokur is gone! I only see water. No buildings, not even the church steeples!"

"Incredible," Nils said.

"I can't believe we got out of there in time," Tryg continued breathlessly. "Thank God we recognized what was going on. All that work around the ocean just saved our lives. I bet the water will go inland a good 12 miles and flood Rip and Varmahlio."

"You're probably right," Nils said. "And if it gets to the thermal craters, we're in for a huge steam bath, or worse — volcanic action!"

As the Norwegians continued east toward the central plains, the pair called home to assure their wives they were safe.

The 12-foot seawall at Reykjavik harbor was no match for the biggest of tsunamis. The downtown flooded with 12 to 15 feet of water, enough to pick up automobiles and other debris to float in a brown sludge for miles. The international airport at Midborg was a disaster scene, with large airplanes smashing into the terminal. The water measured 18 feet on the side of the control tower. City Hall and the Parliament Building were no match for this killer wave either. And the famed Hallgrimskirkja, a cathedral that towers more than 200 feet above the city, was an eerie sight in this new world of brown water. The disaster was over in just a couple minutes. It came without warning because there was no one left in the west coast seaports to sound the alarm.

Tryg emailed the video and photos of the tsunami to his office, where the horrific sights were relayed to national media outlets. Soon the world learned of the tragedy. The grainy videos streamed across the Jumbotron in Times Square. Television programming in America was interrupted to cover the disaster. In Miami, Matt was in his office when he got a call from his wife, Heather.

"Matt, turn on the TV," she said. "It's horrible, just horrible! A tsunami has hit Iceland and whole towns are gone. Thousands have perished!"

"Oh my God!" Matt said as he turned in his chair and remotely activated the television. He called for Murray, Maggie, and some of the Red Box production crew to come into his office. Now there were videos from helicopters on the screen. The Red Box team watched in stunned silence broken only by an occasional, "Look at that!" or "Oh my God!"

Matt looked at Murray and said, "Brother, this disaster will help mobilize the world to do something about global warming. We need to accelerate the moon base program to get the data necessary to make people understand that we must reduce the carbon content in the atmosphere."

"I agree, Matt," Murray replied with a real sense of urgency in his voice. "What's our next action?"

"I'll call California and find out what it will take to get four more Bald Eagles ready for launch," Matt said. "You call Dr. Arnold at NASA to see if he can have more instruments ready in 9 to 10 months."

"Will do, brother," Murray said. "I'll even fly to Huntsville to convene a panel of experts if that's what it takes."

"Good idea, Murray," Matt replied.

Matt then turned to his assistant factory manager, Don Carlos Fuentes. "Don Carlos, please have a plan to increase the Red Box production on this desk by Wednesday morning," he said firmly.

"Maggie, how soon can we leave for Washington?" Matt asked. "I want to meet with the NASA Administrator to get more funds for the necessary changes to Boca Chica and the ISS to accelerate this program."

"I'll take a training T-Pod offline and we can leave by noon tomorrow," Maggie said, sharing her husband's urgency.

"Perfect," Matt replied. "All right, guys. Let's get back to work. Don Carlos, please keep a TV on in the breakroom."

Less than 48 hours later, Matt and Maggie arrived at the NASA headquarters building in southwest Washington, D.C. They were sitting in a seventh-floor conference room when NASA Administrator Golden walked, shook his guests' hands, and said, "Matt, you and your team are always welcome here."

"Thank you, sir. It's a pleasure to be here," Matt responded. "And may I introduce Maggie Flynn, Murray's wife and my pilot?"

"It's a pleasure to meet you, Maggie," Golden said. "Murray's a genius, and it must take one to control him."

Maggie smiled and said, "I try my best!"

The Administrator motioned for everyone to take a seat as he said, "I know why you're here, Matt. I've already had several calls from Dr. Arnold. We can't have another Iceland. They're still finding bodies, and from the seismic data, it looks like the tsunami may have triggered massive volcanic activity. I remember the polar bear that was rolled out with your Bald Eagle spacecraft last year."

"Yes, sir," Matt replied. "I believe you'll have President Werner and Congress' approval when you request additional funds in light of this emergency. Indeed, unless I miss my guess, the President will rally the West, and perhaps even some Asian countries, to step up with funding."

Golden nodded his approval.

Matt continued, "You've committed to launching from Boca to return to the moon and, with industry's help, perhaps even go to Mars. I believe it's time to accelerate the moon base program. The Bald Eagles are perfect to create a cluster on its south pole. However, to accelerate the program, we need to build a garage on the Space Station that can accommodate and replenish four Bald Eagles at one time."

"I hear you, Matt," Golden said. "And I bet you're going to ask to have it done in less than a year. Am I right?"

"Yes, sir. You read my mind," Matt replied.

"Well, guess what?" the administrator said.

"What?" Matt asked.

"We started that planning process six months ago when we got the measurements. We also need it for deep space missions," Golden said. "Indeed, our next launch from Boca will have some of the equipment necessary for a refueling station in the 'garage,' as you call it."

"That's terrific, sir," Matt replied. "And I'm committing to accelerate the production of both the Bald Eagles and the Red Boxes."

"With this commitment, we'll need more astronauts, more mission planners, and more instrument test teams if we're going to pull this off," Golden said. "But Matt, there's one thing we need to do first."

"What's that?" Matt asked.

"The President will always take a call from a White House Medal of Freedom honoree supporting this cause," the Administrator said. "Are you willing to do it?"

"Where's the phone?" Matt said confidently.

"Not so fast," the Administrator replied. "We're working up some preliminary numbers now. Let me put them before him by the end of next week. Then you can call to reinforce *his* idea."

Matt smiled and turned to his sister-in-law, saying, "Maggie, you can take credit for this graduate course idea on how Washington really works."

On the 10ᵗʰ day after the meeting with Golden, Matt and Murray called the President from Matt's office.

"Rita, this is the Flynn brothers," Matt said to the President's appointment secretary. "Could you please put me through to President Werner? It's very urgent."

After a short pause, the Rita responded, "Will do, Matt. He's been expecting your call."

Matt and Murray smiled at each other as the call was put though to Air Force One.

"Hello, Matt and Murray," President Werner said. "I thought you might call."

"Hello, sir," Matt and Murray replied in unison.

"Matt, I'm happy to report we're all over this Iceland tragedy. We've convened a special session at the United Nations to get global support. Our initial Red Cross shipments are already in transit. FEMA will set up a disaster coordination center there. And I'm sending our best seismologists to monitor the underground activity. And, finally, I've just approved a request from NASA to accelerate the moon base program. I hope this makes your day. Now, what was it you wanted to talk about?"

"Mr. President, you *have* made our day," Matt replied.

As he pushed the button on the speakerphone to end the call, Matt said to Murray, "Remember the polar bears. They are on the top of the food chain, but now at our mercy. I hope we're not too late."

CHAPTER 13

RENDEZVOUS II

"Hi, Katia. It's Hans calling," the young astronaut said. "You know next weekend is the long Labor Day weekend, and I'd like to invite you to accompany me to Vance, Alabama."

"Vance, Alabama? What is it? Where is it?" Katia asked.

"It's a small town 18 miles east of Tuscaloosa where the big Mercedes-Benz plant is," Hans replied. "It's at this plant where we make the composite panels for the Bald Eagles. My family owns a small cabin on a lake just south of the plant. There is plenty to do. I'll only have to desert you for three hours as I take a 3-D refresher course. What do you say? Are you ready for some fun away from here?"

"Sure," Katia replied. "It sounds quite interesting. Do I assume correctly that we can take a T-Pod to get some flight time?"

"Right. It's a 90-minute flight, and we'll go right over New Orleans," Hans replied. "I'll pick you up on Saturday at 7 AM. And please be sure to bring your bikini — you know, that orange and white striped one."

Hans pulled up to Katia's apartment building at 6:59 Saturday morning. She was already at the curb with her luggage. Hans got out of the car, hugged her, and put her suitcase in the trunk of his little Mazda. Ten minutes later they pulled up to the terminal at Ellington Field. The couple checked the weather report, filed a flight plan, and completed a walk-around inspection of the T-Pod. A NASA technician inserted a Red Box in its cradle and gave the keys to Hans, reminding him that the Red Box would have to be locked up in Vance. A few minutes later T-Pod 6026 taxied to the helipad, where the engines turned vertically and the craft rose straight up to 100 feet, then turned eastward toward New Orleans.

As they flew over the Mississippi River, Katia exclaimed, "Wow! It's much wider than I had imagined."

"And the city is so low, below sea level, that you can see why it is so prone to flooding," Hans said.

Forty minutes later Hans landed on the helipad at the Mercedes-Benz plant. After the astronaut secured the Red Box in a safe in the main SUV assembly building, he checked out a Mercedes sedan. One of Hans' classmates at the technical institute in Germany volunteered to drive Katia to Tuscaloosa.

"Katia, you'll be in time for the farmers' market. You'll love it. It's one of the best in the South. My class ends at 1. I'll join you at 2 for a late lunch, OK?"

"Sounds great," Katia replied as she warmly accepted a hug from Hans prior to getting in the car.

At 2 PM Hans entered Tuscaloosa and called Katia to make arrangements to pick her up. They found each other and headed to a barbecue stand. Katia couldn't stop talking about all of the neat things she had seen at the market. After lunch the couple toured the Jemison-Van de Graaff mansion and the Tuscaloosa Museum of Art.

"Now what, Mr. Tourmeister?" Katia asked.

Hans had it all planned out. They would go to the cabin, open it and air it out, freshen up, and return to Tuscaloosa for dinner.

They drove south for 6 miles, turned off the main highway, and went 2 more miles on a gravel road to the edge of a forest.

"We're almost there," Hans announced to Katia. "And don't worry — I'm not kidnapping you for some rubles."

They both laughed as they pulled up to a brown wood and fieldstone cabin. At the front door, Hans retrieved the key from its hiding place, opened the door, and motioned for Katia to step inside.

"Oh, Hans, this is really neat," Katia said as she walked around the living room. "A living room, bedroom, bathroom, kitchenette, and a fireplace — it's really all you need. And look at all those fish trophies on the wall. I bet those bass put up quite a fight!"

"You'll find out tomorrow, my dear," Hans said as he checked the stove and the refrigerator, which was filled by a neighbor the day before at Hans' request. He then opened the back porch door and asked Katia to walk down to the lake. They carefully stepped down 50 flagstone steps to the pier, which had a bass boat on one side and a canoe secured at the end.

"Oh, Hans, this is perfect," Katia said as she relaxed in the late afternoon sunshine.

"I'm so glad you think so, honey," Hans said, then blushed. "I'm sorry. I told myself I wouldn't be too forward. It just slipped out." Katia came to Hans, kissed him on the cheek, and said, "You're sweet, too."

Two hours later the couple returned to Tuscaloosa to find their dinner table at the Cypress Inn overlooking the Black Warrior River, directly across from the campus of the University of Alabama. Dinner by candlelight found the two astronauts almost in heaven.

They returned to the cabin around 9:30. Hans started a log fire in the fire pit, which was halfway down to the lake. They changed into jeans and T-shirts, put some mosquito repellant on, grabbed a bottle of Grand Marnier and two glasses, and headed down to the fire pit. The astronauts challenged each other to name constellations in the vast night sky. Hans moved over and sat next to Katia, clinked her glass, and told her of his fondness for her. He could feel an erection gaining stiffness.

It wasn't too long before the couple found themselves undressing each other. Hans found exquisite delight in removing Katia's black lace bra and panties. He ran his hand over her thighs and hardened nipples. He kissed her with a tenderness he had never experienced before. Their warm bodies soon became one, with repeated expressions of a true love several times before sunrise. The smell of frying bacon and fresh-brewed coffee woke up Katia, her nude body chilled slightly by the crisp morning air.

She sat on the side of the bed as Hans brought in a warmed, thick, terrycloth robe and a cup of java. As she slipped into the robe, the goosebumps disappeared. She gave Hans a peck on the check and whispered "I love you" in his ear.

After a joint shower where an innocent bar of soap became a plaything, the couple dressed for breakfast, where they plotted their day of bass-fishing.

"What if we catch one?" Katia asked.

"Well, we have three choices," Hans replied. "We can eat it for dinner, freeze it, or give it to a neighbor. We'll make that decision this afternoon, *if* we catch one."

An hour later they were floating at one of Hans' favorite spots, each with a baited hook in the water.

"Katia, they should be at about 30 feet, so meter out your line to the depth." Hans instructed.

"Will do, Captain," Katia said as she gave him a short salute.

A half-hour passed, and both lines got bites simultaneously. Hans demonstrated the technique as he reeled in his fish. Katia, not to be outdone, was a quick learner and did the same. With two good-size bass on ice and huge smiles on their faces, the boat headed back to the pier.

As they walked up the flagstone steps, Hans said, "These fish will make for a great dinner this evening. We'll roast them right here in the fire pit and baste them with lemon butter. I think you'll like it, honey."

"Sounds great, honey," was her reply.

After a soup and sandwich lunch, the couple applied mosquito repellant and headed down to the end of the pier. They untied the canoe, slipped it into the water, and boarded the craft slowly and carefully, steadied by the pier. They paddled to a secluded site, photographed some great blue herons, and skinny-dipped in the cool, refreshing water.

Happy hour soon arrived and found the couple enjoying a California chardonnay. Katia made a salad while Hans prepared the fish and boiled water for wild rice. Katia hugged Hans from behind as he worked on the fish at the counter.

"No bones, right?" she asked.

"No bones, my dear," he replied.

At 7:30 Hans rang the dinner bell. Katie gave the meal five stars. In fact, she declared it the best fish she'd ever had. "Your second career can be as a chef," she told Hans. They laughed and toasted the wonderful occasion, then began talking of the future.

"What does one do after they have had the thrill of floating around in space?" Katia asked.

"I'm not sure," Hans replied. "Perhaps train the next generation of astronauts. I hope we find out someday."

Hans lit a candle and hinted to Katia that it was time for bed. The flickering light was just enough to highlight Katia's perfectly toned body as the two undressed and jumped under a comforter. Soon they were locked in a lover's embrace, kissing every inch of the other's body. For such a strong woman, Katia amazed herself at how tender she could be. She never felt this way back in the Ukraine. Hans brought out the best of her, and she was truly happy with her unbridled emotion. Hans, on the hand, prided himself with his manliness, but knew that the moment was really the result of Katia's magic touch. They feel asleep in each other's arms.

Early the next day, Labor Day Monday, they closed the cabin and drove back to the plant. With no visitors on Labor Day, Hans gave Katia a private tour of the car museum. She thought she would be bored, but she wasn't. She was particularly impressed with Hans' knowledge of all the safety features Mercedes had pioneered over the past century.

At noon the Red Box was retrieved and inserted into the receptacle on 2026. Katia did an excellent job piloting the T-Pod around a thunderstorm in Louisiana and soon landed at Ellington Field. The couple's ride back to her apartment was filled with fish stories.

As they kissed at Katia's apartment door, she warned, "Watch out, Hans! I'm really loving our relationship!"

Hans smiled as he confessed, "I fell in love with you three months ago during our first rendezvous at Goat Island, Katia. Good night, my love."

CHAPTER 14

MEILAN

While the first Bald Eagles were being manufactured in California, distinguished engineers and scientists came from all parts of China to the Province of Hainan, an island off the southern coast of China. The island, a mecca for tourists flocking to the many five-star resorts in the city of Sanya, with its beautiful bays and beaches, is also famous for the Wenchang Satellite Launch Center (WSLC) on its eastern tip. At this center, the huge Long March rockets are used for moon missions as part of the Chinese Lunar Exploration Program (CLEP).

The meeting was called by Dr. Sun Wei of the Beijing-based People's Aerospace Department. He invited senior aerospace specialists such as Fu Jao Min, Liu Zintao, Wang Yongze, and Sun Jaidon to meet at the Hilton resort hotel in the Meilan section of the northern port city of Haikou. The group spent the evening discussing China's space program, including successes and failures.

On a Sunday morning in a light rain shower, the group boarded a minibus for the 10-minute ride to the Hongcheng Lake district in the center of Haikou. At the same time, three high-ranking military officers from the Central Military Commission were driven by car from the headquarters building of the People's Government of Hainan Province to the lake a quarter-mile away. Near the western shore of the lake is Daokesanki Island, which has only one building, a large, two-story, white stucco conference center reserved for high-ranking government officials.

The minibus and car simultaneously pulled up to the building's parking lot. The officers, engineers, and scientists walked across a short access bridge and entered the building. In the entrance hall, Wei made the introductions and motioned for the group to follow him to a conference room.

Wei took the chair at the head of the table as the others took their seats. There was a teapot at each place. Aides of the officers stood guard outside the door as Wei cautioned his guests that the discussion would be top secret.

After a short introduction about the country's accomplishments in space and praise for several men at the table, Wei introduced Sun Jaidon, who had just returned to his native Hainan after graduating from Cal Tech and spending three years at the Boeing Company as a test technician on the Silent Strike Compact Laser System. While Sun was not privy to many of the engineering aspects of the program, he was very familiar with the components and how to test the system. Sun walked to the end of the table, turned on a laptop, and proceeded to give a very detailed briefing about the laser system. Of particular interest to the group was its tremendous power in a very compact apparatus. Sun showed videos of the laser destroying an unmanned aerial vehicle at a remote site in the Arizona desert.

After a five-course lunch, Wei said, "Gentlemen, we will reverse engineer this device for our own purposes. You are all familiar with the Xichang Satellite Launch Center in southwestern China. Surrounded by mountains, this remote site launches smaller rockets, typically as part of our anti-satellite test program. And as you may remember, back in 2007, an unmanned moon orbiter was successfully launched to begin our lunar exploration program."

Wei continued, "XSLC is home to a complete, state-of-the-art integrations and test facility. At this facility we will build and test a laser system. All of you will be assigned to this important national program for a period of six months. Housing accommodations will be provided. However, no families will be allowed."

After several questions from the group, Wei said, "Our project plan calls for the first prototype test in four months, which is very ambitious. However, this program is of such national importance that it must be done. The three military officers nodded in unison. Colonel Han will be a member of the team. You will start in 10 days. All transportation arrangements will be made for you. Are there any questions?"

The engineers all looked at each other but said nothing.

"Very well," Wei said. "This meeting is adjourned. Please enjoy another night at the hotel as a reward for your accomplishments and your forthcoming efforts on behalf of the People's Republic of China."

CHAPTER 15

CHIN CHIN PO

Four months into her astronaut training, Chin Chin Po had a surprise visitor as she was having lunch in the JSC cafeteria. Matt Flynn came up from behind, tapped her on the shoulder, and said," Hello, Chin Chin. How are you doing?" Chin Chin turned around, jumped out of her seat, and hugged Matt enthusiastically.

"Matt, it's really tough here, but I'm determined to make it!" the pretty Chinese detective said. "The classes are OK, but the swimming is really hard."

"There's no doubt in my mind that you will be a real credit to the astronaut corps!" Matt replied.

"I have great friends here," Chin Chin said as she introduced some of the astronauts with whom she was sitting. Matt sat down with a cup of coffee as the others excused themselves with a friendly, "See you in class!"

After talking about the training for several minutes, Matt said, "I need for you to take three days off and come with me to Washington on important business."

"What kind of business, Matt?" Chin Chin asked. Matt looked around and told her he couldn't discuss it at that location.

Chin Chin recognized Matt's body language and simply said, "Oh."

The next morning a T-Pod took off from Ellington Field with Captain Russell at the controls and Matt and Chin Chin as passengers. Three and half hours later, the craft landed on the military parade field at Fort Belvoir just south of Washington. Captain Russell remained with the craft as Matt and Chin Chin were driven off in big black Suburban.

Only a few minutes later, the vehicle pulled up to the main entrance of the headquarters of the Defense Threat Reduction Agency (DTRA). The huge building, workplace to more than 2,000 employees, plays a key role in destroying weapons of mass destruction. A secondary mission is to reduce the threat of conventional warfare. The two guests were quickly cleared through security and escorted to the fifth-floor conference room. Matt knew what was happening, but Chin Chin was still in the dark.

As she entered the room, she was greeted by the Secretary of Defense, Jefferson Palmer, who said, "Hi, Chin Chin. Thank you for coming."

"Thank you, sir, but I don't think I had much of a choice" Chin Chin said as she looked at Matt. The three of them laughed, and Secretary Palmer motioned for them to have a seat at the table as he began speaking.

"What I am about to say must remain in this room. Is that clear?" Palmer said. His two guests nodded. "Boeing has a compact laser weapon called Silent Strike. A few months ago, it appeared that some of the weapon's digital image files had been inappropriately copied. It was just about the time when two test engineers quit and returned to China. Some of my advisors feel that it is just a coincidence, but my gut tells me that China is trying to reverse-engineer this weapon to create its own."

"What can I do?" Chin Chin asked.

"There is a launch site called Xichang in Sichuan Province," the secretary said. "It is now a military post. Indeed, a few years ago the Chinese successfully launched a kinetic kill rocket, which destroyed a retired weather satellite. We believe this is the most likely place for them to develop a laser weapon. It's my hope that you can call on some of your trusted friends to research this, in complete confidence of course, and get back to me in less than a month."

"Sir, I'll do my best," Chin Chin said. "As fate would have it, one of my former detective colleagues in Hong Kong came from Liangshan, a small town about 40 miles from Xichang."

The secretary wished Chin Chin good luck in the astronaut program. "I've got a feeling you'll be in space before you know it," he said.

Three weeks after the meeting at DTRA, Chin Chin met Matt and Murray at the Boca Chica Control Center.

"Your intelligence is accurate. The device exists and has successfully destroyed ground and airborne targets not far from Liangshan. Some of the evidence is here on this memory stick," she said as she handed a flash drive to Matt. "The laser is encapsulated in a 6-foot metal tube that looks like a telescope."

"Wow! Great work, Chin Chin. Please express our appreciation to you friends!" Matt said.

"You know this is really timely and important," Murray added. "The Chinese have just announced a moon mission launch via the Long March 5 from Wenchang coming up in three months. They call it a robotic mission to study the NEO 2009FD asteroid. This 'telescope' could be deployed on the south pole of the moon, just where we're planning our moon base."

That afternoon Chin Chin, now a qualified T-Pod pilot, flew back to Houston as the Flynn brothers contacted the Pentagon on a secure landline.

CHAPTER 16

GARAGE

"Three, two, one. We have lift-off of Bald Eagle TWO on its journey to the International Space Station," came over the loudspeaker at Boca Chica. Onboard the spacecraft were three mate collars and three astronauts: a 34-year-old commander/pilot, a former fighter pilot from Brazil; a 44-year-old payload specialist, a former Navy Seabee construction engineer; and another payload specialist, a 34-year-old mechanical engineer from Canada. The payload specialists had spent many hours working together in the NBL, practicing the attachment of the three mate collars to the ISS to create the "garage" necessary for three Bald Eagles to be refueled for their trip to the moon. The garage actually had four hatches, so one would always be available to a Dragon spacecraft for resupply or astronaut evacuation in the event of an emergency.

Matt was in the Control Center to witness the launch and attachment videos. Cameras on the ISS and Bald Eagle would provide real-time viewing of the operations.

Twelve minutes after lift-off, Launch Director Preston reported BE-TWO had achieved orbit and was positioning the spacecraft to make the remaining 100-mile journey to the ISS. As the sun set over the Rio Grande River, BE-TWO approached the ISS. Even though the craft was buffeted by high solar winds, the pilot expertly closed the gap after a very anxious 10 minutes, which seemed like an eternity to Matt.

But soon both the pilot and the Control Center confirmed the craft's attachment to the ISS.

During the next two days, the payload specialists were assisted by two ISS crew members in the attachment process, using battery-powered hand tools during six spacewalks. Upon completion of the final mate collar, the Bald Eagle crew was given a tour of the ISS and a send-off dinner of warmed dehydrated space-pack food.

At 10 PM on the third day of the mission, the BE-TWO crew was back aboard the spacecraft, accepting waste canisters and securing them while the pilot went down her prelaunch checklist. By 11:11 the ISS was in the right position over Earth for the separation maneuver. On large screens at the Control Center, Matt could see the spacecraft slowly float way from the ISS. She was on her way home as the Earth glowed blue and white below.

A hundred miles above the Indian Ocean, a micrometeorite penetrated the craft's port-side outer hull and severed an oxygen line. The fast-thinking Seabee quickly crimped the line to prevent any more loss of oxygen. An alternate oxygen line was used during the remainder of the mission.

A water landing was planned in the Gulf of Mexico just off the Texas coast. BE-ONE had already proven it could safely return to the launch pad at Boca Chica. However, in the event that weather would prevent a landing on the pad, the launch team needed a water landing as an alternative for safety. Recovery boats were in place 30 miles from the mouth of the Rio Grande River when BE-TWO passed over North Africa and decelerating through 500 miles per hour via the pilot's adroit use of the Red Box energy control system and the firing of the propulsion module for 60 seconds before being jettisoned. Three miles above the target landing zone coordinates, two parachutes deployed. Splashdown occurred at 11:59 AM. BE-TWO bobbed up and down in the water, a red and white striped parachute floating alongside the craft. Within three minutes scuba divers were securing a floatation collar around BE-TWO.

Shouts of joy went up in the Houston and Boca Control Centers. The obligatory cigars were handed out, though they would not be lit inside the building. Matt leaned over the shoulder of the Propulsion Director and was quite pleased to see that a monitor showed the amount of reserve propellant would have allowed another orbit around Earth prior to splashdown.

That afternoon, Matt received several congratulatory calls, including one from the Pentagon and another from the White House. "Thank you, Mr. President," Matt said. "We are on schedule for our moon mission six months from now."

One week later, Murray inspected the spacecraft in the Integration & Test Facility prior to its trip to Hawthorne for refurbishment. I&T Manager Bob Hock assured Murray, "She's in good shape and will be ready for launch in just five months."

Murray smiled and replied, "I sure hope you're right, Bob."

CHAPTER 17

OUR FAVORITE

In the midst of the celebrations at Boca Chica for the BE-TWO mission, Matt called Chin Chin Po.

"Chin Chin, Captain Russell and I are at Boca," he said. "As you can imagine, it's been quite a day."

"Yes, Matt, I saw the splashdown on the cafeteria screen. Congratulations!" Chin Chin responded excitedly.

"Does your schedule allow for dinner with two old guys tomorrow night?" Matt asked.

Chin Chin checked her iPad and replied, "Yes, I'm free tomorrow night."

"Great," Matt replied. "We'll pick you up at your apartment at 6:30, OK?"

"I look forward to it," Chin Chin said.

The following afternoon, Matt piloted the T-Pod for the short flight from Boca to Houston under the very watchful eye of Captain Russell. Both of them were surprised at the soft landing at Ellington Field.

As they shut down the craft and unlocked the Red Box from its cradle, Matt said to the captain, "It's always fun to see our favorite astronaut!"

"I totally agree, Matt. She has been an asset to our whole program," Jim said.

Promptly at 6:30, Matt's Mercedes-Benz pulled up to the astronaut's apartment complex. He called Chin Chin, and moments later she came down to the car. She was stunning in black silk pants and a red silk matador jacket. As she climbed into the backseat of the car, Matt said, "Chin Chin, you look ravishing and not a day older my dear. What's your secret?"

Chin Chin's face flushed. "It must be in the genes," she said. The trio laughed as Matt turned out of the parking lot and headed north to downtown Houston. Twenty minutes later they entered Brennan's of Houston, ready for some classic Cajun food. Matt had made a reservation for a corner table in the courtyard by the fountain.

As they sat down, Chin Chin told her friends, "I haven't had a date in six months. And, coincidentally, it's my birthday today!"

Matt looked at the Jim and said," Skipper, we're doing something right these days — the mission and landing yesterday, and tonight a date with a beautiful celebrity."

After they placed their drink orders, Matt asked Chin Chin if there was any new news from her sources in China.

"Matt, your timing is impeccable, "she replied. "Just yesterday a convoy was seen leaving XiChang headed for Hainan province. It should be at the Wenchang launch site tomorrow. The speculation is that it has payloads for the Long March moon mission just nine days from now."

"Yup, I think we're on to something here," Matt said. "The Pentagon tracked those two former Boeing test engineers to Xichang before the trail went cold."

"It doesn't take a rocket scientist to see what's happening," Jim said.

"True, my friend. Very true," Matt responded.

After a dinner that lived up to its reputation and a cupcake-and-candle birthday salute, Matt surprised Chin Chin with a comment. "You know, Chin Chin, some of the other astronauts may get jealous, even angry, if you're assigned to one of the early missions," he said.

The Asian beauty was quick to respond. "Well, you have to admit that I've had what you Americans call a 'crash course' on becoming an astronaut," she said. "Fortunately, my math and physics have allowed me to compress the schedule. I've crammed two years into one. That's why you're my first date in six months. And thanks to you, Captain, my flight training has been a breeze and a real joy."

By 9:45, Matt and Jim were hugging Chin Chin good night and wishing her good luck in the future.

Back in Miami a few weeks later, Matt got a call from Chin Chin.

"Guess what?" she said.

"You've been assigned a mission," Matt replied.

"How did you know? Are you behind this, Matt?" she asked with an inquiring tone.

"No, I'm just a cheerleader," Matt replied.

"Well, OK," the happy astronaut said. "I've been assigned as a payload specialist on BE-FOUR for the moon mission this fall.

"That's super. Congratulations!" Matt said. "I'm sure you'll be a valuable crew member. Everyone around here will be thrilled with the news, and we'll watch for you on television."

"I appreciate the support," Chin Chin said before hanging up the phone. "And the next time you're in Houston, the dinner's on me!"

A broad grin came over Matt's face as he ended the call. He immediately telephoned Jim and Murray to relay the good news. Then he asked Murray to come to his office.

As Murray stood in the doorway, Matt said, "Brother, mark my words: You're going to be asked to conduct an important trade study in the very near future by our friends in D.C."

Two weeks later, Murray was in one of his "Einstein modes" as he scribbled formulas on a whiteboard in his office. Only Matt knew what he was up to.

CHAPTER 18

LONG MARCH 5

China's fourth space vehicle launch facility was located on the eastern tip of the Hainan Island province. This 3,000-acre facility at Wenchang, situated just 19 degrees north of the equator, took advantage of the atmosphere of the rotating planet to boost large launch vehicles. The site and its 12 launch pads were ideal for the next-generation heavy lift launch vehicle called Long March 5 (LM5). Modeled after U.S. and Russian rockets, the Long March 5 could launch the equivalent of a loaded school bus into Low Earth Orbit. However, its primary purpose was to put things and people on the moon and perhaps beyond.

These behemoth launch vehicles were made at the port of Tianjin near Beijing and transported by barge to the Wenchang launch site. Much like the space shuttles before them, the vertical three-stage rockets with solid boosters strapped on the side were rolled to the launch pad by a huge transporter, perhaps the largest vehicle on Earth. The whole system had to be huge because, without a Red Box anti-gravity device, the Chinese had to rely on size to counter the pull of gravity.

China launched a communications satellite to the Earth-moon midpoint, called L2, in order to relay commands and data to and from the moon. The country's astronaut corps was eager to plant the red flag on the moon, both on the far side and the southernmost area, usually referred to as the Aiken Basin near the rim of the Shackleton crater.

Aptly enough, one of their designated landing sites in this region was called "Market for Soldiers."

The countdown for the LM5 launch was at T-9 minutes, with a planned hold to check fuel line sensors. A small spectator viewing stand was set up 8 miles away. Western journalists were not allowed on the site, but several press releases described the payloads on the unmanned mission. One payload was a robotic rover with a soil claw to test for minerals which could potentially be mined. Another was a telescope on a triangular base and deployed solar panels for power; its advertised purpose was to study nearby asteroids. A large parachute augmented by retro rockets was designed to land the payloads softly at the south pole.

Fifteen minutes later, the main engines started and slowly lifted the LM5 into the clear, starry night sky. Seismic sensors in Japan and Korea confirmed the launch, as did the USS *Ronald Reagan* aircraft carrier, which was monitoring the disputed zone in the South China Sea where China had constructed artificial islands for landing strips, ports, and military bases.

Six days after the picture-perfect launch, a radio frequency signal expert at the Army's Camp Casey in the northern sector of South Korea asked his sergeant to verify what he just heard and seen on a display.

"Sarge, I think they are activating the telescope. Look at this signal pattern," he said.

The sergeant peered at the squiggly lines on the display and said, "Sure enough. Our intelligence personnel have told us what to look for, and that signal sure looks like the signal to open the lens cover."

He sat down next to the technician and instructed him to focus on signals that would pivot, raise, and lower the instrument. His patience was rewarded just 20 minutes later, when the technician pointed to the display and said, "Yup, there it is, just as we predicted."

The sergeant asked that the data be archived for confirmation by his captain and transmittal via satellite to the Pentagon.

Later that day, Murray got a telephone call. The voice on the other end simply said, "Operation Silent View is a go."

Murray knew just what to do.

CHAPTER 19

NEW TOOL

Maggie took a Red Box and overnight luggage to the Homestead Airport, where she prepared a T-Pod for flight. With everything checked out, she called Murray and said, "We're set to go, my dear." Murray hung up the phone, placed a device in a ballistic satchel, and drove to the airport. The T-Pod knew the way to Houston, making for an uneventful flight. At Ellington Field, Maggie powered down the T-Pod, ejected the Red Box from its cradle, and joined Murray for the seven-minute ride to the NBL.

As the couple signed in at the NBL reception desk, the Director came up and welcomed them with a hearty handshake. "We've been expecting you," he said. "You certainly have friends in high places in order to put this experiment on a fast track for the moon mission. And you deserve it, given the contribution your invention has made to this industry and many others. Welcome!"

"Thank you, sir. You are most kind," Murray said. "Now, if you would be so kind as to lead us to our favorite astronaut."

"Certainly. This way, please," the Director replied as he motioned for the couple to follow him down the hallway. Soon the trio paused at a large window overlooking the huge swimming pool where astronauts practice extra-vehicular activities in the buoyancy of the water.

Soon Murray and Maggie were led into a secure classroom in the basement, where they were immediately greeted by Chin Chin Po, who hugged both visitors. A young man stood next to her. She introduced him as Lennon, a fellow astronaut/energy engineer assigned to the mission.

"You look great, Chin Chin," Murray said as Maggie nodded her approval. "This astronaut life must really agree with you."

"It sure does," Chin Chin replied. "The instructors, contractor, and fellow astronauts here are all fantastic."

"Good," Murray replied. "Now, on to the business at hand. What I'm about to show both of you should be pretty simple to operate, although nothing is simple in space."

As the group sat at a round table in the classroom, Murray placed upon it a large aerosol-type can with a side handle and a spout on top. Chin Chin asked to hold it, and Murray slid it over to her.

As she picked it up, she said, "Wow! This is pretty heavy." She then handed it to Lennon, who agreed with an astonished look on his face.

"It's heavy because it has a lead liner to contain the chemical," Murray said. "But as you know, gravity on the moon is only one-sixth of the gravity here on Earth. It will feel light enough in operation."

Maggie powered up her laptop and showed a series of slides about the Chinese telescope that had just been implanted on the moon. The graphics were simple but had been verified by some of the sources Chin Chin herself trusted.

Murray then turned to Lennon and said, "Sir, you are a British citizen with a Top Secret security clearance. What you are about to hear is considered a national secret by our government. Therefore, you are not allowed to discuss this meeting or anything to do with this device. Is that clear?"

"Yes, sir. I understand," the Brit replied.

"I sincerely hope you do," Murray said, looking the astronaut in the eyes. "If any divulgence of information is attributed to you, you'll be removed from the astronaut corps, extradited back to England, and tried in a military court of law. Do you understand that?"

"Yes, sir. I fully understand the consequences," Lennon said.

Murray nodded to Maggie, who then pulled what looked like a Geiger counter out of a satchel, flipped a switch, and walked around the room, often pointing under chairs. As she came back to Murray's side, she said, "All clear."

"Good. Now we can proceed," Murray said. He went over to a whiteboard and used a laser pointer to make a point.

"You see, the telescope is not a telescope at all," he began. "It's a laser weapon thousands of times more powerful than this little pointer I have in my hand."

Maggie showed a slide of the Aiken Basin area, where the United States planned to establish its moon base.

Murray pointed to the screen and said, "This is the approximate location of the weapon, which according to good intelligence is pointed at the very location where we plan to establish our base. Your mission will be to use this device to neutralize the weapon. And rest assured your activities will not be televised." Murray picked up his invention and aimed it at the screen for emphasis.

"The moon's big," Chin Chin said. "What if we don't land near the target?"

"Good question," Murray replied. "Tomorrow you will have two sessions in the NBL. Sitting on the bottom are two space-qualified Segway transporters. We've even created a 12-foot-by-12-foot rocky area where you will develop your skills. A rover is too large to be attached to the spacecraft, and these compact transporters can be secured inside the craft. The rest of the crew will believe you two are conducting a mining experiment, prospecting for Helium-3 and titanium."

"Tomorrow morning at 9:30 you will enter the NBL with the assistance of three Navy frogmen," he continued. "The area will be cleared of all other personnel. These frogmen will assist you in your training. Do you have any questions?"

There was silence as the meeting closed. As the two astronauts left the room, Chin Chin turned around and winked at Murray. He gave her the thumbs-up sign.

The next morning the T-Pod left JSC and flew southwest to Boca Chica, where Bob Hock gave Murray and Maggie a personal tour of the I&T Facility, where six Bald Eagles were undergoing their final tests.

At 3:30 the couple again boarded the T-Pod, this time heading to New Orleans for a much-anticipated anniversary dinner at Arnaud's and the bedtime frolicking that was sure to follow.

CHAPTER 20

LIFT-OFF

With the successful test launches of the Bald Eagle spacecraft and the garage at the ISS completed, the authorities at NASA finalized their plans for establishing the moon base. Given the various international funding sources and the diversity of the crews, MoonBase 1 would be an international research station and not the property of a single nation. It would exist only for peaceful purposes, with the primary mission to help solve the global warming crisis on Earth and the secondary mission to support a Mars mission.

Because the Red Box eliminated the launch vehicle and the Bald Eagle design accommodated rapid refurbishment, the Moon mission was possible only 18 months after first successful landing at Boca Chica. The mission was divided into two phases. In Phase 1, two unmanned spacecraft — LOGPOD and LABPOD — would be sent to the edge of the Shackleton crater via remote control. The tricky part would be to land them close enough to be able to deploy the mate collar joining the two Bald Eagles. Phase 2 would have two Bald Eagles transport nine astronauts to the ISS, where they would be replenished with expendables before continuing to the south pole of the moon. The LOGPOD and the LABPOD would remain on the moon to service future missions. The two CREWPODS would return to the ISS during Day 6 of the mission and separate from the ISS on Day 8 for vertical landings at Boca.

On a crisp October morning, BE-THREE, the LOGPOD, rolled out of the I&T Facility and took its position on the launch pad. This Bald Eagle was critical to the success of the mission. It contained the electric power and life-sustaining gases essential for a moon-base cluster of Bald Eagles. Aboard the LOGPOD were battery packs, backup fuel cells, oxygen tanks, and extra filters because the 6 inches of moon dust gets into everything.

The television coverage began a day before the launch. Global excitement was expressed by crowds from Auckland to Sao Paulo and everywhere in between. Space agency officials from around the world were treated as VIPs and enjoyed Texan hospitality. Matt Flynn, NASA officials, and a busload of engineers from Hawthorne were in attendance. Murray, who had been feeling under the weather, stayed in Miami, nursed by Maggie, the love of his life for 26 years.

The countdown began with Matt in the Control Room VIP suite. He carried the Red Box handcuffed to his wrist. Upon the signal from Launch Director Preston, he left the building with an entourage of technicians and went to the scaffolding that surrounded the spacecraft. Matt climbed up the stairs, released the handcuffs, and witnessed the insertion of the Red Box in its cradle. The LOGPOD was powered up after personnel and handling equipment were removed from the launch pad. All systems were nominal as Matt returned to the Control Center and shook the hand of the astronaut-pilot who would guide the spacecraft through space and land it near the crater.

"May your mission be a success, sir," Matt told the pilot. "May you help save the polar bears!" Everyone around them knew what he meant.

The countdown resumed with only 20 minutes before launch. Matt called his brother and said, "Murray, you worked day and night to help devise the software to make this robotic mission possible. I love you, brother!"

There was a pause, then Murray replied, "Just remember, brother, almost everything that happens in the space program is done by the lowest bidder!" A laugh and a tear was shed on both ends of the telephone line.

Millions of people around the world anxiously watched the television as the countdown reached "one." The Bald Eagle lifted straight up into the azure blue sky. When it reached an altitude of 1,000 feet, one of the aerozine engines fired on queue. Soon the craft was only a speck on top of a vapor trail. It would take two days to travel the 243,000 miles to the moon. It was the highlight of the evening news around the planet.

The next day Matt went to Houston to watch the landing in the Control Center, where he stood next to Chin Chin in the visitor's room. He was there only 10 minutes when he heard the Launch Communicator say, "We are go for touchdown." Matt hugged Chin Chin for good luck and visualized Murray, Maggie, and Heather watching the event on television back in Miami.

As the descent camera showed BE-THREE slowly descending to the moon's surface, Matt thought of the importance of Chin Chin's role in the upcoming mission. Composure was difficult, but somehow he managed to stand erect to hear, "The Bald Eagle has landed." Thunderous applause engulfed the room as everyone either hugged or shook the hand of those around them.

In a quiet voice, Matt asked Chin Chin, "Are you ready, my dear?"

Chin Chin looked him in the eye and said, "To me my mission is bigger than all of China."

He hugged her again.

With the successful landing, the deployment of the solar arrays, proven recharging of the battery packs, and deployment of communications antennas, the mission was a total success. As a result, the prelaunch activities for the BE-SIX or LABPOD were initiated.

Nine days later, with the cigar smell finally gone from the VIP room of the Boca Control Center, the LABPOD was rolled from the I&T Facility to the launch pad. The LABPOD was important to the success of the moon base because it contained many experiments focused on growing plants in the lighter, oxygen-free atmosphere of the moon. One of the most important experiments was the small prototype O'Neill cylinder, with its alternating windows and land stripes that contained seeds and small hydroponic plants.

Just as before, the countdown was held at 20 minutes as the Launch Director made his final checks. It seemed longer, but 20 minutes later, the Launch Communicator counted down, "Three, two, one, and we have lift-off." Thanks to the Red Box, the LABPOD ascended straight up 1,000 feet before the aerozine motor fired to put the craft on the proper trajectory to the moon. The LABPOD, with all of its experiments and the medical dispensary, was the heaviest of the Bald Eagles that made up the moon base. Therefore, more engine thrust bursts were required to leave Earth.

Controllers in Houston and Boca Chica were on high alert as the LABPOD fired its side-mounted Drago jets to lower the craft onto the moon's surface an incredible 2 meters from the LOGPOD. The mate collar was deployed and locked into place. Matt, who was in Houston again, had a pocket full of cigars, which were quickly handed out.

Two weeks later, two crews took T-Pods from Houston to Boca to prepare for their launch. The BE-ONE crew consisted of Commander Crowley, Katia, Hans, and Colonel Porter, the Marine Riverine pilot. The crew of BE-FOUR consisted of the Brazilian pilot, Canadian payload specialist, the Navy Seabee, and astronauts Lennon and Chin Chin. Each astronaut was eager to prove the worth of his or her training.

Though ready, each was mindful of the fact that mistakes are usually unforgiving in space. Lennon and Chin Chin checked their Segways and "Murray Devices" without the knowledge of the other crew members.

On the morning of the launches, BE-ONE was rolled out to the launch pad. At T-2 hours, the Red Boxes were installed and final leak checks completed. At T-1 hour, the crew arrived and was secured in their seats. The hatch was closed and checked for leaks. The scaffolding was rolled to the safe area. The grandstands had been doubled in size and every seat was taken. Dozens of media representatives were asking thousands of questions. Jim Russell flew Matt and Heather to Boca, and Maggie flew Murray in another T-Pod. No sooner did they arrive in the VIP Lounge when, true to his promise, President Werner arrived.

"Great to see the Flynns here for such a momentous occasion," the President said as he shook the inventors' hands and hugged their wives. He was in a very cordial mood, and even asked Jim if he would return to Camp David for another round of golf.

"Sir, my clubs are in the T-Pod," the Captain responded. "Just name the date."

"Fine, Jim, you're on. We'll do it in a couple months when it warms up," the President said as he made the motion of swinging a golf club.

President Werner than excused himself to hold a press conference in an adjoining room. In a short five-minute speech, he eloquently described the importance of the day and "our new era of space exploration, which will bring jobs to America and new technologies to help the environment." He then pointed to the Flynns, who were sitting in the front row, and said, "I have it on good authority that the polar bears are watching this mission with keen interest." The audience knew it was a salute to Matt and Murray, who had initially helped fund the Polar Bear Xpress moon mission. The Flynn brothers stood up and saluted the President and the audience.

Twelve minutes later the Launch Director approved the final countdown. Mission Control in Houston then issued its approval to proceed. Right on schedule, BE-ONE lifted straight up into the cloudless day on its way to the ISS, where it docked five hours later. The Commander reported, "Houston, Boca, we have a secure lock at the ISS." As the President left Boca, he informed Matt that Secretary of Defense Palmer would represent him at the next day's launch.

At 6:15 PM the following day, BE-FOUR, the final Bald Eagle to compose MoonBase1, lifted off the Boca Chica pad with the glowing, setting sun as a backdrop. As the spacecraft approached the ISS at midnight, the crew marveled at the size the structure. This time it was a female voice that broadcast, "Houston, Boca, we have a secure lock at the ISS."

Cheers could be heard all over south Texas and the rest of the world, although perhaps not as loud in China. The Secretary of Defense shook Matt and Murray's hands as he quietly said, "Operation Silent View is now in motion."

CHAPTER 21

MOON BASE

A day after BE-FOUR docked at the ISS, both BE-ONE and BE-FOUR were cleared for separation from the ISS. BE-ONE, the command ship, confirmed separation with "Moon base, here we come!" To avoid a possible collision, BE-FOUR followed one hour later. With their helmet visors down, the other astronauts could not see the excited smile on Chin Chin's face.

The supercomputers accurately calculated a path through space debris, so both spacecraft approached the moon without incident. After circling the moon at an altitude of 200 miles, BE-ONE was cleared for landing. The flight computer activated the trajectory toward the south pole. At the 12-mile, go-no-go altitude, the spacecraft flipped over and the retro-rockets began to fire. At 3 miles above the surface, the commander and pilot could see the LABPOD and LOGPOD on the surface. The pilot checked his gauges, particularly the fuel gauges, as he adroitly lowered BE-ONE to the surface, a mere 4 meters from the LABPOD in the proper location to form a square when BE-FOUR joined them. No space launch is routine, but somehow this seemed too simple to the millions of people around the world who watched the historic event on television.

The night before touchdown, Murray complained of chest pains and was rushed to the Houston Heart Clinic. With the other Flynns at his side, along with a cadre of doctors and nurses, the group reined in its emotion as it watched the television. Murray was told not to get too excited, and he complied. The cardiac monitor showed only a slight elevation in heart rate as Murray kept his cool. He smiled as Maggie patted him on the head.

All eyes were glued to the television when it was announced, "BE-FOUR, you are go for descent." A female voice replied: "Roger, Houston and Boca, we are cleared for descent." As BE-FOUR reached the 200-mile milestone, BE-ONE reported the successful deployment of the mate collar to the LABPOD. The 12-mile milestone passed quickly, with the retro-rockets firing flawlessly. BE-FOUR was gently lowered into the exact position to form the moon base. Cameras on BE-ONE and the LABPOD recorded the event for posterity. Once the two mate collars were in place — one to BE-ONE and the other to the LABPOD — the moon base could operate as designed, with astronaut movement enabled between the four spacecraft.

Day 1 was devoted to collecting rock samples, analyzing them in the LABPOD, and downloading the data to several institutions. The LOGPOD's solar panels were wiped free of moon dust to improve their efficiency.

On Day 2, Hans and Katia deployed a telescope and instrument cluster that studied lightning strikes on Earth. On a moonwalk that afternoon, the two astronauts set up the base for the O'Neill cylinder. The cylinder concept caught the imagination of both scientists and the media. Consequently, the television cameras focused on the pair's efforts only 30 meters from the moon base.

The cylinder activities were a perfect diversion from BE-FOUR, where Lennon unhooked the mini-Segways and deployed them on the far side of the spacecraft. Chin Chin retrieved the Murray Device from its locker, checked the battery pack, and joined Lennon at the transporters. Lennon gave Chin Chin the thumbs-up sign and signaled for radio silence. The pair mounted their transporters and headed due west at a slow but steady pace. In 16 minutes they arrived at the Chinese telescope, which was indeed pointed directly at the moon base. Lennon secured the two transporters to the surface while Chin Chin videotaped the Chinese writing on a metal plate on the outside of the telescope. Lennon then exchanged the camera for the Murray Device so Chin Chin could perform three tasks. First, she moved the trigger of battery-operated device to the first position. This activated a laser, which she used to melt the communications antenna's feed horn. Second, she moved the trigger to the second position to activate a more powerful laser to cut a hole in the telescope's outer casing directly over the control computer. In seconds the printed circuit board was melted around the controller chip, exactly where Murray had predicted its location would be. Chin Chin than bounced around to the front of the telescope, where she soldered the lens hinge in place. With these actions, it would be impossible to command and operate the laser weapon as intended. Chin Chin gave Lennon the thumbs-up sign as they mounted their transporters for the short trip back to the moon base. With the transporters and Murray Device secured in a canister on the LOGPOD, Lennon opened a secure communications channel and broadcast a cryptic message: "Operation Silent View completed." Only officials at the Pentagon knew what it meant. Three pings on Chin Chin's headset indicated confirmation. The two astronauts then returned to BE-FOUR in time for a scheduled television broadcast titled, "Life on the Moon."

The astronauts spent the rest of the day securing samples in canisters, wiping the omnipresent moon dust off everything, checking expendable supplies, and finishing the televised sessions as the Commander and the two pilots went down their prelaunch checklist. At 6:30 PM, the commander was told by Mission Control, "BE-ONE is cleared for ascent at 8:30." The commander and pilot both acknowledged the communication with, "That's affirmative, Houston. We copy." At 7:15 PM BE-FOUR was informed of its ascent time window, 45 minutes after BE-ONE's launch. The Red Boxes were not activated for this portion of the mission because the moon's gravitational pull is only one-sixth that of Earth's. Both Bald Eagles lifted off on schedule and without incident. At the midpoint between the moon and the ISS, BE-ONE's communications were temporarily interrupted by a violent solar storm. BE-FOUR was then told to dock first at the ISS. Within an hour, BE-ONE was also securely locked in the garage. Both crews assisted with the unloading of some specimens for analysis by the ISS astronauts. Waste canisters were also unloaded, to be retrieved by Bald Eagles on subsequent missions. The crews were treated to videos from both the telescope and the O'Neill cylinder. They rested six hours and alternatively spent a half-hour on exercise cycles before the Commander informed them that the launch window to return to Earth would open in two hours. He announced, "Crews will return to their stations by 7:26 for prelaunch activities." Chin Chin said a silent prayer that this would not be her last venture into space.

In the predawn darkness, caravans of RVs and vehicles of all types started to fill up Highway 4 toward Boca Chica. The spectator stands had been enlarged during the week of the moon mission. Maggie pushed Murray, who was confined to a wheelchair due to heart trouble, as they joined Matt and Heather in the VIP room at Boca. Mimosa cocktails and a Texas-size buffet awaited the guests. The Flynns held sessions with the media and signed autographs for the "spaceniks" in attendance. Matt leaned over and asked Murray, "Brother, how do you feel about these landings? What does your gut tell you?"

Murray paused before he answered, "The Lord has given me the strength and joy to be here. He won't disappoint me."

Matt smiled as he patted his brother on the shoulder and thought to himself how odd it was that Murray suddenly mentioned the Lord. He was not a religious person.

The VIP room quickly filled with politicians, each seeming to take credit for the moon base. A dozen key engineers and their families also were treated to the event by the Bald Eagle manufacturer.

The launch site communicator soon announced that both Bald Eagles had separated from the ISS and were on their way back to Earth. At the 200-mile altitude milestone, the lower section of the spacecraft was ejected and the Red Box activated to let the anti-gravity device provide a controlled descent through the Earth's atmosphere.

With BE-ONE entering Earth's atmosphere, Murray stood up at the window of the VIP room to watch the live video feeds as the craft came over the Atlantic Ocean and started its final descent to Boca Chica. At the same time, Matt thought of his trusted friend, Chin Chin, and hoped her craft would have a successful landing. Heather seemed to know what was on Matt's mind as she came over to him and held his hand, saying, "My love, this will end without incident." Matt always trusted her instincts.

Long before BE-ONE was visible to the grandstands, the long-range cameras saw it flying toward the Texas coast. The launch communicator then broadcast, "In 3 minutes and 39 seconds we will have touchdown." Two minutes before touchdown, the announcement came, "Please turn attention to the area directly above the launch pad. Right on cue, the commercial spacecraft built by the lowest bidder began its final maneuvers with the eight Drago thrusters spewing red flames on all sides of the spacecraft. The communicator counted down the feet to touchdown — "50, 40, 30, 20, 10, and touchdown!" There wasn't a dry eye that witnessed the feat in person. This event was repeated 26 minutes later, when BE-FOUR followed the same scenario.

As the spacecraft stood next to each other, scaffolding towers surrounded them to allow the astronauts to exit their crafts. One by one they appeared in the hatch, waved to the crowd, and walked over to the elevator with assistance from launch personnel. Matt knew which astronaut would be the last to appear, and Chin Chin did not disappoint him. With eight other astronauts on the platform, Chin Chin appeared at the hatch of BE-FOUR and flashed the thumbs-up sign. The crown went wild. But only Matt and Murray and the Secretary of Defense knew what this thumbs-up really meant.

The President called to congratulate the astronauts, mission controllers, launch personnel, and support contractors for the achievement. He said, "I look forward to future missions to the moon base and the science that will help solve our environmental problems here on Earth."

Quietly the Flynns excused themselves from the celebratory barbecue and strapped themselves into their T-Pods for the trip to Miami. Four hours later the twin Bald Eagles landed at the Homestead General Aviation Airport. Large Mercedes-Benz Maybach limousines awaited and took the couples to their respective residences.

Matt then called Chairman von Boltz to thank him for his support during the entire program. The Chairman thanked Matt and said, "You Flynns deserve the world's gratitude. You have largely solved the transportation problems on Earth while making the universe more accessible. I look forward to our next venture together."

As Matt hung up the phone, he thought about the executive's words. Somehow he had read Matt's mind: "Thank God for the peaceful uses of space."

CHAPTER 22

WHAT HAPPENED?

Right after the telescope went silent, the Chinese officials again met on Daokesanli Island in central Haikou. The same group of engineers and scientists were present when the meeting convened. One by one the senior scientists were asked by Wei, "What happened? Why did our instrument go silent?" Liu Zintao had no answer. Neither did Wang Youngzi. The young engineer, Sun Jardon, who helped reverse-engineer Boeing's invention, offered an opinion as he said, "Sir, the design was correct. It worked perfectly in XiChang. We had the right mount so launch vibration or sound would not destroy it. And we had a soft landing at the correct location on the moon."

"So, what is your theory?" Wei asked.

"Sir, I believe it's been sabotaged by American astronauts on the recent moon mission," Sun responded.

"Can you prove it?" the high-ranking official asked.

"No, sir. I can't. But it seems possible to me. Somehow the Americans knew it wasn't a telescope and disabled it during a moon walk just about the time it went silent."

"Engineer Sun, that is our belief as well," Wei responded. "What do I tell the United Nations officials who have initiated an inquiry sponsored by the British? I am due to meet with them next week in New York City."

"Sir," Sun replied, "I'm an engineer, not a politician. I can only promise you a replacement in nine months. That is, if all of the subcontractors are still in business."

"On my authority, Dr. Sun, you must proceed at once. Do I make myself clear?" Wei asked.

"Yes, sir. Perfectly clear," Sun replied. "We will have an instrument ready for the Long March 5 launch next May."

"Very good. Now, gentlemen, you are excused to begin your important work," Wei said as he closed the meeting.

Early the following week, Wei and a small staff flew to New York to appear before the Committee on the Peaceful Uses of Outer Space (COPUOS). The Chairman, a senior British diplomat, welcomed Wei with a caution: "Sir, as your country well knows, there will be no weapons launched into space. Does China abide by this global policy?"

"Yes, sir. It does," Wei said.

"Was the telescope you recently landed on the moon indeed a telescope?" the Chairman asked.

"Sir, for some reason it stopped working," Wei responded, evading the question. "We're currently investigating the cause while we build a replacement. I hope this meets with universal approval."

There was no objection in the room. However, after the meeting, a cryptic message appeared on Secretary Palmer's iPad. It read, "Round Two has begun."

He knew what it meant for world peace.

CHAPTER 23

SANYA

The island province of Hainan lies at China's southernmost point. On the eastern coast of Hainan is the Wenchang Launch Complex, where the LM5 was launched earlier in the year. On the south coast of the island is the resort city of Sanya. In this semitropical environment, with its many bays and beaches, an amazing array of five-star resorts has been developed in the past 10 years.

Five months after returning to Earth, astronauts Lennon Blair and Chin Chin Po were asked to attend the 52nd Global Astrophysics Conference at the Ritz-Carlton Resort in Sanya. The conference was dominated by American astrophysicists describing incredible images from the Hubble Space Telescope, but Lennon and Chin Chin's mission was to hear firsthand what the Chinese were saying about their telescope on the moon.

The first two days were very relaxing for the pair, with massages, spa treatments, and lounging by the swimming pools. Occasionally they would attend a seminar, but as Lennon put it, "It's way over my head!"

On the morning of the third day, when a famous Chinese astrophysicist was due to give a paper, the lecture was dropped from the program. Members of the Western media immediately began asking why and wondering aloud where the images are.

Coerced into providing an explanation, a Chinese government spokesperson told the audience that the telescope had suddenly stopped working. In the entire audience there were only two people who knew that it wasn't really a telescope.

From the back row of the auditorium, Lennon and Chinn Chin captured the announcement on their cell phones and quietly left to pursue what Lennon had earlier promised Chin Chin. An hour later the two astronauts were at a golf driving range. Chin Chin looked incredible in a yellow polo shirt and white Bermuda shorts. After Lennon described the purpose of each club in the bag, he demonstrated the proper position to hit the ball. It was then Chin Chin's turn. She was a quick learner, but Lennon couldn't resist standing behind her with his arms around her, demonstrating the proper swing. After a few misses and few partial hits, Chin Chin hit the ball the correct distance for the type of club she held in her hand. Lennon came over to her and gave a congratulatory high-five.

Soon they were on the putting green. Here, too, Lennon was all too eager to put his arms around his willing pupil. At that moment he thought of Miko, his friend and fellow astronaut back in Houston. He really liked her, but it was primarily for her physical attractiveness and aggressive nature. With his arms around Chin Chin, it was different. It was both physical and emotional. He was smitten by this smart, beautiful woman who had come into his life.

He was determined to keep her there.

CHAPTER 24

CEREMONY

Three days later, Lennon and Chin Chin landed at George Bush Intercontinental Airport. Because cross-town traffic in Houston can be brutal, they were glad to get a ride in Matt's T-Pod south to Ellington Field. As she left the T-Pod, Chin Chin asked, "Matt, are you and Captain Russell going to accept my dinner invitation?"

Matt looked at Jim, and they both smiled and nodded. She then turned to Lennon and invited him too.

"Sure, Chin Chin. I'd be delighted," Lennon replied.

"Great. We're on for 6:30," Chin Chin said. "Let's meet at my apartment, and we can drive in one car from there."

The group members then went their separate ways and reconvened promptly at 6:30.

Chin Chin was a regal hostess at a steakhouse in downtown Houston. As they finished dessert and coffee, Matt said he had an email to share with them. "Here on my cell phone is a message for both of you," he said. "It says, 'Please come to the White House on July 26 for dinner and an award.'"

Chin Chin's jaw dropped as she looked around the table.

"Chin Chin and Lennon, this is no joke. It's for real," Matt said. "Jim and I will be there, as will my brother and our wives. This will be very special."

"What a surprise!" Chin Chin said as she broke out in broad smile. "Matt, I bet you're involved in this."

"Well, I did have a vote, yes," Matt replied.

As the party said goodnight back at Chin Chin's apartment, Matt and Jim drove away and saw Lennon embracing Chin Chin in the rearview mirror.

The guests arrived at The Willard InterContinental hotel two blocks east of the White House in the afternoon of July 26. Maggie brought Murray in a T-Pod. Captain Russell piloted another with his wife, Jackie, and Matt and Heather. Chin Chin and Lennon arrived via airline and were picked up by a limousine.

Promptly at 7, two black Cadillacs escorted by two black Suburban SUVs picked up the guests for the two-minute ride to the north portico of the White House. The women wore evening tuxedos. The men also wore tuxedos, and Matt and Murray both wore their Presidential Medals of Freedom.

The President and Mrs. Werner warmly greeted the party at the front door and escorted the guests into a reception room. Secretary of Defense Jefferson Palmer and his wife walked across the room and greeted the party. The NASA Administrator then introduced himself and his wife. Lennon and Chin Chin beamed with anticipation. As a toast was being enjoyed, Secretary Palmer pulled the NASA Administrator aside to inform him of the special mission the two awardees had performed while on the surface of the moon. Murray was very quiet and seemed distant at times. Matt, on the other hand, lacked no enthusiasm for the occasion and confidently led many conversations.

Upon completion of the six-course dinner, the President asked Secretary Palmer and the two astronauts to join him at the head of the table. At that moment a staff person brought in two flat, blue-velvet boxes and laid them on the table. The President opened one and gave the other to Secretary Palmer. Each box contained a plaque and a very special medal bearing an eagle and dangling from a maroon, blue, and white ribbon. The President asked Secretary Palmer to say a few words.

"It is my distinct honor to present each of you the Secretary of Defense's Medal for Distinguished Public Service. This is the highest award I can make to a foreign national, and I do so with heartfelt gratitude for your achievements as a valued member of the astronaut corps. With this award you join the ranks of several past presidents."

Secretary Palmer then tied the medal around Lennon's neck and read the citation on the plaque. Then President Werner said a few words about Chin Chin's past services on behalf of the Red Box as he tied the medal from behind.

As the two recipients accepted the applause of the dinner guests, the gold medals reflected the light of the room's chandeliers. At that point in the ceremony, the President asked Matt to say a few words and also make an award. Quite eloquently, the senior Flynn congratulated the brave astronauts, the Bald Eagle technologies, and the recently completed moon mission that captivated the world.

Matt then asked the guests to rise and join him over at the window side of the room facing the south lawn. As he took his place next to Chin Chin, he said, "Chin Chin Po, please accept this gift as a token of our gratitude for all that you have done for America."

Matt then nodded to a staff member to draw the curtains, at which time a spotlight illuminated the south lawn. Directly in the middle of the lawn was a gleaming white T-Pod with a large Bald Eagle decal on the side. At that moment, Murray walked up to Chin Chin and said, "Here are the keys, my dear."

The room erupted in applause as the staffers and kitchen personnel poured into the room from all directions. Chin Chin leaned on Lennon and wiped away tears of joy.

At that point NASA Administrator Golden tapped on a glass. After the applause died down, he said, "Lennon and Chin Chin, I'm granting you both a month's paid leave. You've earned it. Please enjoy it with the heartfelt gratitude of our nation."

Jim and Jackie Russell escorted the honorees back to the hotel as the Flynns remained behind for post-ceremonial beverages in the library with President and Mrs. Werner. It was also the President's way of bouncing ideas off two of his most trusted friends, the Flynn brothers.

The next day the Flynns took separate T-Pods back to Miami. Chin Chin and Lennon were the guests of Matt and Heather for a day while Chin's Chin's T-Pod was delivered to the Homestead airport.

The next morning the two were driven to the airport. An hour later they were airborne to fulfill a special request from Murray to fly to Ambergris Caye off the coast of Belize in Central America. The purpose was to enjoy the very hotel that was the headquarters during the development of the Baldie seaplane, the first application of the Red Box. They weren't about to refuse the assignment.

Indeed, the couple made good use of honeymoon suite at the Sun Breeze resort. The bartender there introduced the two travelers to the Baldie, a civilized drink comprised of two parts Orangina and one part Bacardi Gold rum.

As Lennon kissed Chin Chin for the nth time, he asked out loud, "How could life be this sweet?"

CHAPTER 25

GUESS WHAT?

Murray was deep in thought in front of the whiteboard in his Miami office when NASA's Chief Scientist, Dr. Cliff Arnold, called. Murray slowly went over to his desk and picked up the phone.

"Hello, Dr. Arnold," he said. "Glad to hear from you. How can I help you?"

"Guess what?" Dr. Arnold replied.

"What?" Murray asked.

"You know that O'Neill cylinder experiment we left on the moon?" Dr. Arnold said. "Well, it's small, only 10 feet long, but it's working. I'll send you a time-lapsed video that shows vegetables growing."

"That's great," Murray replied. "But as I remember, O'Neill's concept called for tubes to be 20 miles long."

"That's correct, Murray," the doctor said. "I'll get to the point. We intend to build one a mile long with a diameter of 1,500 feet. Habitats have failed in the past, and we don't want to repeat a failed experiment. We've got a location in mind about 40 miles west of Albuquerque, New Mexico."

"Interesting," Murray said. "How can I help?"

"I would like you to run some calculations about the atmospheric content necessary to sustain both human and plant life," Dr. Arnold explained. "I have a supercomputer you can use. What do you say, Murray? Can you spare some time to help us?"

"Doctor, you caught me at a good time. I'm between self-imposed projects, and your project sounds challenging," Murray replied in an upbeat tone.

"Great!" the doctor said. "I've just emailed a video and some background to you. Please view it and apply your smarts to the problem. I'll get the contractual paperwork done by the end of the week."

"Fine," Murray answered. "Thanks for your call and the new challenge."

As soon as he put the phone down, Murray went into his now famous "Einstein mode."

Three weeks later, Murray emailed his initial calculations to NASA, which spurred a return call by Dr. Arnold the following day.

"Fascinating, Murray," the astrophysicist began. "If I read your data correctly, the mile-long cylinder would sustain a crew of eight for almost a year, perhaps longer depending upon the carbon dioxide absorption rate of the plants."

"That's correct," Murray said. "Please use the data as you see fit. But remember, the selection of the plants will be critical. Now, if you'll excuse me, doctor, I'm feeling tired and missing my nap."

Dr. Arnold thanked Murray and ended the conversation by saying, "Take a long nap, Murray. You've earned it!"

Murray slumped back in his chair.

Three months later, Murray was invited to the ground-breaking ceremony for the O'Neill cylinder at Laguna, New Mexico. Maggie piloted the T-Pod over the two-day trip with an overnight stay in Houston to watch a Bald Eagle launch at Boca Chica. The next day the T-Pod flew over a parched west Texas and landed at the Albuquerque International Sunport to sit out a summer squall before proceeding to the Laguna pueblo region. The NASA Administrator, Golden, and Dr. Arnold greeted the couple with open arms.

"So glad you could make it," Dr. Arnold said.

"I'm really pleased to be here," Murray replied. He patted Maggie on her back and said, "We took our time, and I was in good hands with my able pilot here." Everyone smiled.

During the ceremony, Administrator Golden recognized Murray's contributions to the space program and asked him to stand and take a bow. Murray obliged and waved to the grandstand.

The Administrator closed his remarks by saying, "A year from now we'll return to cut the ribbon on this new, exciting test facility called the O'Neill Cylinder Experiment, or OCE. It should foster man's plans to have sustainable habitats here on Earth and elsewhere in the universe. Dr. O'Neill would be very proud of this initiative."

On the return flight back to Miami, the T-Pod again stopped in Houston, where the couple enjoyed a lovely dinner with Chin Chin and Lennon. It was Lennon who asked Murray, "Guess what?"

"What?" Murray replied.

"Guess who put up most of the money for the OCE?" Lennon asked.

Murray had a quizzical look on his face. Lennon paused, then smiled and said, "Your brother."

Murray just shook his head and said, "I should have known. He's trying to save the planet by himself."

CHAPTER 26

BALD EAGLE SOARS

After the Laguna trip, Murray's health deteriorated. His heart was giving out and he was losing the will to live. Confined to his house, he lounged by the pool and had very few visitors. Maggie stopped her Red Box testing duties to be with him full time. She was the gate through which one had to pass in order to have a short conversation with her beloved genius.

On the afternoon of September 12, Matt swung by Murray's house on his way home from the Red Box factory. Maggie greeted him at the front door and escorted him poolside to see his brother. Maggie propped her husband up in a chair with a pillow. Matt updated his brother about the daily events, including that a souped-up T-Pod could now go 350 mph. Murray's eyes rolled as he beckoned Matt to lean over his chair so he could say something. Matt did so and Murray asked, "We're saving the polar bears, aren't we, Matt?"

Murray had drawn his last breath. Matt looked up with tears in his eyes and said, "Maggie! Maggie, he's gone!"

Maggie felt faint. She slumped by the chair and uttered, "Oh, my God. My precious Murray!" Matt put his arms around her as they sobbed together.

Three days later Murray was cremated per his wishes. Condolences poured in from all over the world, even from China. A memorial service was held at the Van Orsdel Funeral Chapel in South Miami. There was a seemingly endless caravan of black Cadillacs and Suburban SUVs filled with government officials who wanted to pay their respects. Astronauts Chin Chin, Lennon, Hans, and Katia were given bereavement leave to attend the service. Jim Russell gave a very moving eulogy, leaving no dry eye in the room. Matt listed some of the accomplishments of his brother and their impact on mankind for centuries to come. Maggie was stoic and refused to cry in public.

Matt commissioned life-size bronze statues of Murray to be made and placed in parks in Key Biscayne and Homestead, Florida, and near the Mall in Washington, D.C. SpaceX funded the fourth statue for placement at the gate of Boca Chica. Street names were changed in Murray's honor around the world.

A week after the memorial service, Matt was poolside with Heather and Maggie. "You know," he said, "we still have one of my brother's wishes to fulfill. He wants his ashes scattered over the Gulf of Mexico. Let's do it and continue on to Ambergris Caye for a vacation."

"Lord, I sure need it," Maggie quickly responded. And let's invite Captain Jim and Jackie to join us."

"Good idea, Maggie," Heather said.

It only took a week to close out most of Murray's affairs. On a bright Sunday morning, with Jim as the pilot and a crew of four, a yellow T-Pod lifted off the tarmac at Homestead airport and headed west over the Everglades. Just as the craft was turning south toward the Gulf, Heather pointed out her window and said, "Look!"

On the port side at the same altitude and speed was a magnificent bald eagle. Matt looked at Maggie. Their eyes met as they both seemed to say, "That could be Murray in another life, still flying with us."

www.ingramcontent.com/pod-product-compliance
Lightning Source LLC
Chambersburg PA
CBHW030814180526
45163CB00003B/1287